W9-AQN-779

NEWTON

on the Continent

ALSO BY HENRY GUERLAC

Science in Western Civilization, a syllabus
Lavoisier—The Crucial Year
Essays and Papers in the History of Modern Science

NEWTON
on the Continent

HENRY GUERLAC

Cornell University Press

ITHACA AND LONDON

THIS BOOK HAS BEEN PUBLISHED WITH THE
AID OF A GRANT FROM THE HULL MEMORIAL
PUBLICATION FUND OF CORNELL UNIVERSITY.

First published 1981 by Cornell University Press.
Published in the United Kingdom
by Cornell University Press Ltd.,
Ely House, 37 Dover Street, London W1X 4HQ.

International Standard Book Number 0-8014-1409-1
Library of Congress Catalog Card Number 81-3187
Printed in the United States of America
*Librarians: Library of Congress cataloging information
appears on the last page of the book.*

For MIELLE TENNEY *and*
JUSTINE OLIVIA BATTERSBY

Preface

My title, I realize, can easily mislead or disappoint the unwary reader. Isaac Newton was immune to the lure of travel. He never crossed the Channel, as did many of his English contemporaries. Even his migrations in England were limited to Lincolnshire, to Cambridge, and eventually to London, whence, on one occasion, he undertook a daring visit, doubtless by boat, to Greenwich to see the astronomer John Flamsteed, one of his less fervent admirers. Oxford, it would appear, he never troubled to see.

The greater portion of this collection constitutes what I have elsewhere called a "reputation study," in this case two longish papers dealing with Newton's *fortuna* on the Continent, notably France, in Newton's lifetime the European center of scientific activity. These papers are preceded by short studies pointing to echoes of European thought to be found in Newton's earliest surviving notebook.

Two of the papers that make up the present modest book were written after the publication of my *Essays and Papers in the History of Modern Science* (The Johns Hopkins University Press, 1977). Two of the earlier articles have been substantially altered since they first appeared, one in *The British Journal for the History of Science* (1968) as a review of two volumes of *The Correspondence of Henry Oldenburg*, the other ("Amicus Plato and Other Friends") in *The Journal of the History of Ideas* (1978). The concluding paper, published here for the first time, is the offshoot of a project that has engaged me for some years: the preparation of a critical edition of Newton's *Opticks*. Intended to form part of the Introduction to that forthcoming work, it soon outran its original purpose, and I have made it a major chapter of this collection. My work on this concluding paper and on the revision of the two others was aided by a fellowship awarded me in 1978–1979 by the John Simon Guggenheim Memorial Foundation and by some research support from my university. I owe special thanks to my friend Alain Brieux of Paris for providing me with the engraved portrait of Pierre Varignon.

HENRY GUERLAC

Ithaca, New York

Contents

Illustrations

Abbreviations and Short Titles

Birch, *History of the Royal Society*	Thomas Birch. *History of the Royal Society of London.* 4 vols. London, 1756–1757.
DSB	*Dictionary of Scientific Biography.* 16 vols. Charles Coulston Gillispie, editor in chief. New York, 1970–1980.
Hist. Acad. Sci. (or Mém. Acad. Sci.)	*Histoire et Mémoires de l'Académie des Sciences* I, 1666–1686. Paris, 1733. The *Histoire* and the *Mémoires* are separately paginated, and are cited separately.
Koyré and Cohen (1972)	*Isaac Newton's Philosophiae Naturalis Principia Mathematica.* 2 vols. Ed. by A. Koyré and I. B. Cohen. Cambridge, Mass., 1972.
Motte-Cajori	*Sir Isaac Newton's Mathematical Principles of Natural Philosophy and His System of the World.* Trans. by Andrew Motte in 1729 and ed. by Florian Cajori. Berkeley, Calif., 1934.
Newton *Correspondence*	*The Correspondence of Isaac Newton.* 7 vols. Cambridge, 1959–1977. The first three vols. were edited by H. W. Turnbull; vol. 4 by J. F. Scott; and vols. 5–7 by A. Rupert Hall and Laura Tilling.
Newton's Papers	I. Bernard Cohen et al. *Isaac Newton's Papers and Letters on Natural Philosophy.* Cambridge, Mass., 1978.
Oeuvres de Huygens	*Oeuvres complètes de Christiaan Huygens.* 22 vols. The Hague, 1888–1950.
Oldenburg *Correspondence*	*The Correspondence of Henry Oldenburg.* 10 vols. Ed. and trans. by A. Rupert Hall and Marie Boas Hall. Madison and London, 1965–1975.

Phil. Trans.	*Philosophical Transactions of the Royal Society of London.*
Principia (1687)	Isaac Newton. *Philosophiae Naturalis Principia Mathematica.* London, 1687.
Whiteside, *Newton's Lectures* (1973)	*The Unpublished First Version of Isaac Newton's Cambridge Lectures on Optics, 1670–1672.* Cambridge: The University Library, 1973. A facsimile of MS Add. 4002, with an introduction by D. T. Whiteside.
Whiteside, *Newton's Mathematical Papers*	*The Mathematical Papers of Isaac Newton.* 7 vols. Ed. by D. T. Whiteside. Cambridge, 1967–1976.

NEWTON
on the Continent

CHAPTER 1

Amicus Plato and Other Friends

In a valuable article, Richard S. Westfall made a careful study of Newton's earliest student notebook, a notebook begun when Newton entered Cambridge in 1661.[1] A good portion of it consists of notes Newton took from Aristotle's *Organon* and *Ethics*, Johannes Magirus's *Physics* (an abridgement of Peripatetic physical thought), and other works a student was supposed to master. About the year 1664, however, Newton began to use the notebook for a series of entries headed *Quaestiones quaedam philosophicae* in which he set down reflections inspired by his own self-propelled reading of Descartes, perhaps Gassendi, and certainly Gassendi's English disciple, Walter Charleton, as well as those familiar influences upon him: Robert Boyle and the Cambridge Platonist, Henry More.[2]

At the top of the first page of these *Quaestiones* Newton had written, probably at some later date, what Westfall calls "a slogan." It reads, without punctuation: *Amicus Plato amicus Aristoteles magis amica veritas*. Westfall passes this by without lingering, and merely comments that "Whatever the truth in the following pages, Plato and Aristotle do not again appear. Newton had indeed been meeting new friends."[3] We are left to assume, if we wish, that this "slo-

---

Revised, with permission, from *Journal of the History of Ideas*, 39 (1978), copyright © 1978.

[1]Richard S. Westfall, "The Foundations of Newton's Philosophy of Nature," *The British Journal for the History of Science*, 1 (1962), 171–182. This notebook was discovered, and its contents briefly summarized, by A. R. Hall in "Sir Isaac Newton's Note-book, 1661–65," *Cambridge Historical Journal*, 9 (1948), 239–250.

[2]Cambridge University Library, MS. Add. 3996.

[3]Westfall, "Foundations," 172.

gan" was of Newton's devising. Yet this is not the case, and several years later Westfall identified the seventeenth-century source from which Newton borrowed his "slogan," without, however, calling attention to its venerable lineage.

Newton, it is clear, had adapted to his purpose a proverb well known in the later Middle Ages and the Renaissance,[4] although commonly found in a version (or versions) differing from that given in his notebook. Yet its import was always the same: to assert aphoristically the primacy of truth over authority or friendship, a good example, despite a variable cast of characters, of what Arthur Lovejoy called a "unit idea." In this note I shall point to some changes the proverb underwent, hinting thereby at a kind of skeletal or encapsulated history of an idea.

In the various early occurrences with which scholars are familiar we often find the name of Socrates substituted for that of Plato, and the name of Aristotle does not appear at all. Yet the aphorism has sometimes been associated with Aristotle himself; and in this there is more than a grain of truth: for while it does not appear verbatim in any of Aristotle's works, the proverb is a succinct paraphrase of a passage in the *Nicomachean Ethics*. Aristotle, discussing the Idea of Form of the universal Good, remarks that the inquiry "is made an uphill one by the fact that the Forms have been introduced by friends of our own. Yet it would perhaps be thought to be better, indeed to be our duty, for the sake of maintaining the truth even to destroy what touches us closely, especially as we are philosophers or lovers of wisdom; *for while both are dear, piety requires us to honor truth above our friends.*"[5]

Aristotle is, in fact, indebted to Plato, the inventor of the Forms, for the idea expressed in that passage. In the *Republic* (10.595) Plato has Socrates arguing that poetry is dangerous to the state, and remarking: "Although I have always from my earliest youth had an awe and love of Homer . . . for he is the great captain and

---

[4]E. Margalits, *Florilegium proverbiorum universae Latinitatis* (Budapest, 1895), 31, col. 2.

[5]The translation is in W. D. Ross, *Works of Aristotle Translated into English*, 12 vols. (Oxford, 1908–25), IX, *Ethica Nicomachea* 1. 5. 1096a. 13–17. Ross's translation was used by Richard McKeon, *The Basic Works of Aristotle* (New York, 1941), p. 939.

teacher of the whole of that charming tragic company; but a man is not to be reverenced more than the truth, and therefore I will speak out."[6] Again, in the *Phaedo,* Socrates says: "This is the state of mind, Simmias and Cebes, in which I approach the argument. And I would ask you to be thinking of the truth and not of Socrates."[7]

But clearly we are far from having the sentiment compressed into an aphorism. For some time, the earliest postclassical aphoristic statement I was aware of was the following passage from Roger Bacon's *Opus Majus* (Part I, cap. 7) given here in my translation: "For Plato says, 'Socrates, my master, is my friend but a greater friend is truth.' And Aristotle says that he prefers to be in accord with the truth, than with the friendship of our master, Plato. These things are clear from the Life of Aristotle and from the first book of Ethics and from the book of secrets."[8]

Bacon's younger contemporary, Saint Thomas Aquinas, also gives us the substance of our aphorism. In his commentary on the *Nicomachean Ethics* he writes:

> For Andronicus the Peripatetic says that sanctity is that which makes men faithful and observant of those things relating to God. And close to this is the judgment of Plato who, rejecting the opinion of his master Socrates, said that one should cherish the truth more than anything else. And elsewhere he says, Socrates is indeed a friend, but a greater friend is truth. In another place, he says one must care little for Socrates, but much for truth.[9]

[6]*The Dialogues of Plato, translated into English by B. Jowett, with an Introduction by Raphael Demos,* 2 vols. (New York, 1937), I, 852.

[7]Ibid., p. 475.

[8]"Nam Plato dicit, 'Amicus est Socrates, magister meus, sed magis est amica veritas.' Et Aristotelis dicit, 'se magis velle consentire veritati, quam amicitiae Platonis, doctoris nostri.' Haec ex vita Aristotelis et primo Ethicorum, et libro secretorum, manifesta sunt." *The 'Opus Majus' of Roger Bacon,* ed. John Henry Bridges, 2 vols. (Oxford, 1897), I, 16. Cf. *The Opus Majus of Roger Bacon,* translation Robert Belle Burke, 2 vols. (Philadelphia, 1928), I, 17–18. It should be recalled that Roger knew the *Phaedo,* which had been rendered into Latin ca. 1156, along with the *Meno,* by Aristippus of Catania. See the *Opus Majus,* ed. Bridges, II, 274.

[9]*S. Thomae Aquinatis in decem libros ethicorum Aristotelis ad Nicomachum expositio,* ed. R. M. Spiazzi (Turin and Rome, 1949), Lib. I., Lectio VI, ¶ 78. My confrère on the Board of Editors of the *Journal of the History of Ideas,* John F. Callahan, pointed out to me, after my article appeared in that journal, that this important occurence had escaped me.

It is from Bacon, however, that we learn something of the path by which our proverb reached Europe. His reference to a book of secrets is doubtless to that pseudo-Aristotelian compilation, the *Secretum secretorum,* translated into Latin from the Arabic in the twelfth or early thirteenth century, and a great favorite of Bacon's.[10] Unless it somehow escaped my notice, our proverb does not appear in Bacon's own version of the *Secretum,* as edited by Robert Steele,[11] or in Gaster's translation of the Hebrew version.[12] Yet it may well be found with, or in, some of the many MSS of this immensely popular work, a work which one scholar has innocently entitled in English *The Privy of Privies.*

The *Vita Aristotelis* that Bacon mentions deserves a word of explanation. Some compilers of dictionaries of quotations and proverbs cite our aphorism in the form "Socrates is dear to me, but dearer still is truth,"[13] and the reader is referred to a life of Aristotle attributed to Ammonius, son of Hermias, as printed by A. Westermann as an appendage to a nineteenth-century edition of Diogenes Laertius's *Lives of the Philosophers.*[14] As Westermann informs us in his Preface, this *Vita Aristotelis* was first printed by Aldus with Ammonius's commentary on Aristotle's *Categories* in 1503. But it has an interesting background which deserves a quick review.

Our proverb appears in differing forms in a *Life of Aristotle* found in three distinct medieval manuscripts, two Greek and one Latin.

[10]See Lynn Thorndike, *A History of Magic and Experimental Science,* 8 vols. (New York, 1923–1958), II, 267–278.

[11]*Opera hactenus inedita Rogeri Baconi, Fasc. V. Secretum Secretorum,* ed. Robert Steele (Oxford, 1920).

[12]"The Hebrew Version of the 'Secretum Secretorum'" translated by M. Gaster, *Journal of the Royal Asiatic Society* (1908), 111–162.

[13]See Francis H. King, *Classical and Foreign Quotations* (New York: Frederick Ungar, n.d.), p. 14, and Georg Büchmann, *Geflügelte Worte,* 32nd ed. (Berlin, 1972), pp. 509–510. The 1925 edition of Büchmann (p. 366) urges the reader to note the opposite sentiment in Cicero's *Tusculan Disputations,* given in German as "Lieber will ich mit Plato irren, als mit jenen (den Pythagoreern) das Wahre denken." This remark is omitted in the later edition of 1972.

[14]*Diogenes Laertii de clarorum philosophorum vitis,* ed. C. Gabr. Cobet. *Accedunt Olympiodori, Ammonii, Iamblichi, Porphyrii et aliorum vitae Platonis, Aristotelis, Pythagorae, Plotini et Isidori* Ant. Westermann. (Paris, 1878). The Greek text reads (p. 10, col. 1, lines 42–43): φίλος μὲν Σωκράτης, ἀλλὰ φιλτέρα ἡ ἀλήθεια, and is Latinized as *Amicus quidem Socrates sed magis amica veritas.* Westermann's additional material is separately paginated.

The so-called *Vita Marciana,* a much mutilated single manuscript in the Biblioteca Nazionale di San Marco in Venice, written about 1300, was first edited by L. Robbe in 1861[15] and in supposedly improved form by V. Rose in 1886.[16] More recently it has been reproduced by Ingemar Düring and by Olof Gigon.[17]

The second Greek version, called the *Vita vulgata,* is commonly found in MSS of the text of Aristotle's *Categories,* or with commentaries on that work. According to Düring, it is more obviously Byzantine, with "many examples of vulgar Greek idiom."[18] It is found in some thirty-one manuscripts, of which Düring believes nine are earlier than 1300. This is the text that was appended by Westermann to Cobet's edition of Diogenes Laertius, and attributed to Ammonius, the Neo-Platonic commentator of Aristotle, a pupil of Proclus, and the teacher of Philoponus, Simplicius, and others in his school at Alexandria.

Finally, existing in sixty-five MSS, there is the *Vita Latina.* Düring writes: "The Latin *Vita* was translated from a Greek original, very similar to but not identical with *Vita Marciana,* probably about A.D. 1200."[19]

All of these versions, it appears, go back to a Greek "Life of Aristotle" written by a certain Ptolemy. The identity, and of course the *floruit* and base of operations, of this Ptolemy have not been determined; the name, after all, was common in Alexandrian culture. We are told that epitomes of Ptolemy's Life of Aristotle were used in oral teaching in the school of Ammonius Hermiae, and by his disciples, including Olympiodorus and Philoponus. Düring estimates that three generations of students from about 480 A.D. until the middle of the following century were taught the epitome of Ptolemy's Life of Aristotle. The versions we have described are said

---

[15]L. Robbe, *Vita Aristotelis ex codice Marciano Graece* (Leiden, 1861).

[16]*Aristotelis Fragmenta* (Leipzig, 1886), pp. 426–450.

[17]I. Düring, *Aristotle in the Ancient Biographical Tradition* (Göteborg, 1957) and *Vita Aristotelis Marciana. Herausgegeben und kommentiert von Olof Gigon* (Berlin, 1962) in *Kleine Texte für Vorlesungen und Ubungen,* ed. Hans Lietzmann. I was steered to these useful sources by my friend Professor Harold Cherniss.

[18]Düring, p. 137.

[19]Ibid., p. 144.

to have descended from epitomes studied in the school of Ammonius. This, and the fact that these derivative lives often accompanied Ammonius's commentaries—for example his commentary on Aristotle's *Categories* printed by Aldus in 1503—account for the ascription of the authorship to Ammonius printed (yet questioned in his Preface) by Westermann.[20]

What of our proverb? In the Greek *Vita Marciana* our tag is given in one form only: "It is Plato who said we should have little concern for Socrates but very much for the truth."[21]

The *Vita vulgata*, on the other hand, gives three versions of the quotation:

(1) For he [Plato] said that one should care much more for the truth than for anyone.

(2) Socrates is my friend, but a much greater friend is the truth.

(3) We should have little concern for Socrates, but very much for the truth.[22]

The *Vita Latina* here echoes the *Vita vulgata*, rather than the *Vita Marciana*, for we read: "For it is a saying of Plato that one should care more for truth than for anything else. And elsewhere he says: 'Socrates is indeed my friend, but a better friend is truth.' And in another place: 'Indeed we must care little for Socrates, but a great deal for truth.'"[23]

Judging from the medieval examples cited, Plato is considered as the originator of our proverb, and the Socrates of the *Phaedo* appears as the friend who suffers by comparison with the truth. But I do not find that, until the Renaissance, Plato replaces Socrates as the friend in question. There is however, a possible exception. In a recent article, Manfred Ullmann of Tübingen has reported his discovery in an Arabic text of a citation that is earlier than those I have cited. It is from the physician, Rufus of Ephesus, who flourished under Trajan (ca. 100 A.D.) and who is quoted as

---

[20]*Diogenes Laertii de clarorum philosophorum*, etc., ed. Cobet, iii.

[21]Düring, pp. 101–102; Gigon, p. 4, lines 129–131.

[22]Düring, p. 132.

[23]"Platonis enim est sermo quod magis oportet de veritate curare quam de aliquo alio. Et alibi dicit 'Amicus quidem Socrates, sed magis amica veritas.' Et in alio loco 'De Socrate quidem parum est curandum, de veritate vero multum.'" Düring, p. 154.

writing: "A philosopher ordered that one should give children [only?] water to drink; I cannot accept this opinion, for truth is dearer to me than any person."[24] The implication, according to Professor Ullmann, is that Plato is the philosopher whom Rufus does not mention by name.

It would be of especial interest to discover the form that the aphorism assumed when it is cited during the sixteenth and early seventeenth centuries. One might expect that under the influence of such an array of Aristotelian critics as Lorenzo Valla, Luis Vives, Ramus, and Gassendi, someone might have been impelled to add the name of Aristotle to that of Plato when invoking this venerable proverb. Ramus, in fact, in his *Defensio pro Aristotele adversus Jac. Schecium* (Lausanne, 1571), makes use of the proverb in a free version of its Socrates-Plato form.[25]

I cannot claim to have made an exhaustive search, but what I have turned up is both interesting and, with respect to a mention of Aristotle, negative. An early sixteenth-century example is from Luther's *De servo arbitrio* of 1525, where the aphorism takes the form: *Amicus Plato, Amicus Socrates, sed praehonoranda veritas.*[26] Here Plato has joined Socrates as an authority. A likely source offered itself in the *Adages* of Erasmus, but this brought no results. My inspection of those editions of the *Adages* published in his lifetime, those at least which happen to be in the Cornell University Library, showed no citation of our proverb. Margaret Mann Phillips, our chief authority on the *Adages,* confirmed my findings, offering the explanation that Erasmus confined his selection to quotations of explicit classical pedigree.[27]

But a related path brought an interesting result. After the death

---

[24]Manfred Ullmann, "Die Schrift des Rufus 'De infantium curatione' und das Problem der Autorenlemmata in den 'Collectiones medicae' des Oreibasios," *Medizin historisches Journal,* Band 10, Heft 3 (1975), 165–183. I owe my awareness of this article to my friend Dr. Owsei Temkin.

[25]Vives in his *Satellitium vel Symbola (Opera Omnia,* ed. Majans, IV, 32–64) does not include the proverb in his small collection, but does give (No. 90, p. 46) the related and more familiar apothegm: *Veritas, temporis filia.* If truth is indeed the daughter of time, it is superior to the doctrines of any one man. The Vives and Ramus references I owe to my wife, Rita Guerlac.

[26]Büchmann, *Geflügelteworte* (1925), p. 366, and (1972), p. 509.

[27]Personal communications, letters of 12 Oct. 1976 and 16 July 1977.

of Erasmus other humanist scholars took it upon themselves, or were urged by the publishers, to make additions to Erasmus's collection, which indeed had already grown steadily in the later editions of the *Adages* produced by Erasmus himself.[28] One such posthumous edition, published in Paris in 1579 with comments by Henri Estienne, has additional adages derived from such lesser humanists as Polydore Vergil, Charles de Bovilles, John Hartung, Adrien Turnebus, and others.[29] One group of contributions is introduced with the running title *Adagiorum Ioannis Vlpii Franekerensis Frisii.* Who Joannes Ulpius was I do not know, except what his title tells us, namely that he was evidently a native of Franeker in Friesland. It is he, in any case, who adds to this edition of the *Adages* of Erasmus what seems to have become the common form: *Amicus Plato, magis amica veritas,* followed by the same proverb in Greek, and then by several Latin lines explaining the meaning of the proverb. This in turn is followed by the single word "Galenus." A search of the index volume of the Kühn edition of Galen brought no results. When I appealed to Dr. Owsei Temkin, who knows his Galen as well as anyone, he could not recall encountering the aphorism anywhere in the Galenic corpus; but he remarked that the dictum "would be quite in line with Galen's emphasis on the primacy of truth."[30]

By the end of the sixteenth century Plato sometimes takes Socrates's place in the proverb, although the two are still occasionally linked together as they had been by Luther, for example by Michael Maier in the "Epistola" of his *Symbola Aureae Mensae.*[31] But

---

[28]For the growth of the *Adages* from the *Adagiorum Collectanea* of 1500, through the *Chiliades* of 1515, to Erasmus's death in 1536, see Margaret Mann Phillips, *The 'Adages' of Erasmus: A Study with Translations* (Cambridge, 1964), Part I, pp. 3–135.

[29]*Adagiorum des Erasmiroterodami chiliades quatuor cum sesquicenturia: magna cum diligentia, maturoque iudicio emendatae & expurgatae. Quibus adiectae sunt Henrici Stephani animadversiones* (Paris, 1579). See cols. 1225–1226: *Adagiorum Ioannis Vlpii Franekerensis Frisii Epitome.*

[30]Personal communication, letter of Dr. Owsei Temkin, 17 July 1977.

[31]Michael Maier, *Symbola aureae mensae duodecim nationum* (Frankfurt, 1617) where the Epistola Dedicatoria (unpaginated) gives the proverb as *Amicus Socrates, Amicus Plato, veritas magis amica.* See also the letter of Maximos Margunios (10 August 1598) to David Hoeschel, in Polychronis K. Enepekides, "Maximos Margunios an Deutsche und Italienische Humanisten," *Jahrbuch der Österreichischen Byzantini-*

what seems on closer examination to be the favored form appears, of all places, in *Don Quixote* where Cervantes cites the proverb just as we encountered it in the posthumous *Adages* of Erasmus. The knight writes an avuncular letter to Sancho Panza who has become governor of the island of Barataria, and concludes: "I owe more to my Profession than to Complaisance; and as the saying is, *Amicus Plato, sed magis amica Veritas*. I send you this scrap of Latin, flattering myself that since thou cam'st to be a Governor, thou may'st have learned something of that Language."[32]

This at least brings us to the early seventeenth century, for Part I of *Don Quixote* was published in 1605; and Part II, where our Latin tag appears, came ten years later, and became available in English through T. Shelton's translation of 1620. Not until later in the century—so far as I know—does the form of our proverb encountered in Newton's student notebook of 1661–1665 make its appearance.

The earliest known occurrence of our proverb in the Aristotle form was turned up, in the course of a different sort of inquiry, by Samuel Eliot Morison nearly a half century ago. He noted it on the title page of the *Philosophemata*, a collection of essays by the Calvinist divine and casuist William Ames (1576–1633), published posthumously in 1643. Here it reads *Amicus Plato, Amicus Aristoteles, sed Magis Amica Veritas*.[33] The inclusion of the motto on the title page

---

schen Gesellschaft, 10 (1961), 122, and G. Podskalsky, *Theologie und Philosophie in Byzanz* (Byzantinisches Archive, Heft 15, 1977), p. 87. I owe these references to Dr. Karin Figala of Munich.

[32]Quoted from Ozell's revision of the English translation of Peter Motteux in *The Ingenious Gentleman Don Quixote de la Mancha by Miguel de Cervantes* (New York, 1950). The annotated editions of *Don Quixote*, investigated by my colleague Professor Ciriaco Arroyo, give no printed source for Cervantes's quotation. Instead they cite, after Nuñez de Guzman's *Refranes o proverbios en Romance* (1555 and later eds.) "un refran nuestro: 'Amigo Pedro, amigo Juan, pero mas amiga la verdad.'" An interesting metamorphosis, indeed, of Plato and Socrates! See, e.g., the critical edition of *Don Quixote* by Francisco Rodriguez Marin, 7 vols. (Madrid, 1927–1928), VI, 68, n. 8.

[33]Samuel Eliot Morison, "Harvard Seals and Arms," *The Harvard Graduates' Magazine*, 42 (1933), 1–15. Morison cited the Ames title page to argue that it was the source of Harvard's motto (*Veritas*) and, with becoming generosity, Yale's (*Lux et Veritas*). The popularity of the proverb, and the innumerable occurrences of the word "truth" in philosophic works, theological treatises, to say nothing of the Bible,

need not have been the choice of Ames himself. For as McKerrow has pointed out, from the last quarter of the sixteenth through the seventeenth century a title page need not be regarded as the production of the author (or the editor of a posthumous work) "but rather as an explanatory label affixed to the book by the printer or publisher."[34]

Perhaps surprisingly (since Aristotle's authority is obviously called in question) our apothegm is cited, not once, but twice, by the natural philosopher Honoré (or Honoratus) Fabri (1607–1688). In the *Dialogi physici* (1669), a work owned by Newton (and which he mentioned in his "Hypothesis explaining the Properties of Light" in 1675), Fabri gives the tag in the form Newton entered in his notebook, not once but twice in the course of the book. As the first Dialogue (*De lumine*) opens, the second interlocutor remarks: "I have nothing to offer in reply except that well-worn proverb *Amicus Plato, Amicus Aristoteles, sed magis amica veritas*, by the love and need for which all those, at least, who ardently pursue letters should be influenced. Whence I should advise you, my friend Alithophile, to value it against your adversaries."[35]

Since Newton did not become familiar with Fabri's book until long after he had ceased to make entries in his *Quaestiones*, it was probably not Fabri who first introduced him to that common proverb (*tritam illam paroemium*). It is much more likely, as Westfall suggested in 1971,[36] that Newton's source was the English physi-

---

make this hard to accept despite the circumstantial evidence adduced by that most distinguished of historians. My attention was called to the Morison note and to the Ames motto by a recent paper of Professor Mason Hammond of Harvard where in a long appendix he discusses the proverb and makes extensive use of the early version of this study. He suggests corrections on a number of minor points. See his "Latin, Greek, and Hebrew Inscriptions on and in Harvard Buildings. Part I: Memorial Hall," *Harvard Library Bulletin*, 28 (1980), 299–346. See especially Appendix 3, pp. 344–346.

[34]Ronald B. McKerrow, *An Introduction to Bibliography for Literary Students* (Oxford, The Clarendon Press, 1928), p. 91.

[35]Honoratus Fabri, *Dialogi physici* (Lyon, 1669), pp. 1–2. Newton's copy is now in the Wren Library of Trinity College, Cambridge. This work should not be confused with another work of Fabri's having a similar title, but covering different subjects, and published in 1665.

[36]Richard S. Westfall, *Force in Newton's Physics* (London & New York, 1971), p. 325.

cian, natural philosopher, and original Fellow of the Royal Society, Walter Charleton (1620–1707). Soon after midcentury, perhaps under the influence of Hobbes, Charleton came under the spell of the pioneers of the New Philosophy, notably Descartes and even more obviously Pierre Gassendi. Charleton's *Physiologia Epicuro-Gassendo-Charltoniana or a Fabrick of Science Natural Upon the Hypothesis of Atoms* (1654) played an important role in spreading the gospel of Gassendist atomism in England.[37] It is here, in the first chapter, where Charleton enumerates the various philosophical sects, that Newton doubtless encountered the proverb. There are those, Charleton wrote, "whose brests being filled with true Promethean fire, and their minds of a more generous temper, scorn to submit to the dishonourable tyranny of that Usurper, Autority, and will admit of no Monarchy in Philosophy, besides that of Truth." These men, he continues, "ponder the Reasons of all, but the Reputation of None." Among these Assertors of Philosophical Liberty, as he calls them, he lists Tycho Brahe, "the subtle Kepler, the most acute Galilaeus . . . the most perspicacious Harvey, and the Epitome of all, Des Cartes." He suggests that to honor each of the heroes "we could wish (if the constitution of our Times would bear it) a Colossus of Gold were erected at the publick charge of Students; and under each this inscription: *Amicus Plato, amicus Aristoteles, magis amica veritas.*"[38]

That Charleton was young Newton's source for the line scribbled in his notebook hardly admits of doubt. But although Charleton may have been the first to use the Plato-Aristotle form of the proverb—the name of Aristotle does not accompany Socrates or Plato in the earlier examples I have cited—others may have anticipated him. At all events, it echoed the mood of the century that saw the great philosophic revolution, just as the substitution of Plato for Socrates reflected the Platonist enthusiasm of the Renaissance. Surely we can understand how Newton, largely liberated from the

[37]See Robert Kargon, *Atomism in England from Hariot to Newton* (Oxford, 1966), chap. VIII; and his sketch of Charleton in DSB, III (1971), 208–210.

[38]*Physiologia Epicuro-Gassendo-Charltoniana* (London, 1654), p. 3. A facsimile of this work was issued by the Johnson Reprint Corporation in 1966 (*The Sources of Science,* No. 31) with an introduction by Robert Kargon.

undergraduate grind, heavy with Aristotle and late Peripatetic doctrine, and who had just discovered Descartes, Gassendi, and the rest of his "new friends," would find the device a compact expression of his new freedom and of the alluring vistas he had glimpsed.

CHAPTER 2

Newton's First Comet

I saac Newton's obsession—the word is hardly too strong—with comets will have been noted by anyone even casually familiar with his *Mathematical Principles of Natural Philosophy*, commonly known by its abbreviated Latin title, the *Principia*. In the first edition (1687) of this famous work the treatment of comets occupies nearly a third of Book III, that concluding section where Newton applies his law of gravity to deriving the observed motions of the planets, the moon, and the tides of the ocean. Comets, he made clear, when they are visible to us, appear to circle the sun in paths that could be represented by conic section and, with a very good approximation, by the properties of a parabola. In later editions (1713, 1726) comets take up still more space in this concluding Book. Indeed the need for revisions was in part stimulated by Newton's recognition that he had not solved to his satisfaction the problem of cometary trajectories.[1]

From the first, it was not only the mathematics of the motion of comets that occupied Newton, but the physical nature of those mysterious bodies. Comets, he knew, shine by the light of the sun and are most brilliant when close to perihelion. Their tails, always

---

Slightly expanded from my review of *The Correspondence of Henry Oldenburg,* ed. and trans. by A. Rupert Hall and Marie Boas Hall, II (1663–1665), III (1666–1667), which appeared in *The British Journal for the History of Science,* 4 (1968), and is reprinted here with permission.

[1] I. Bernard Cohen, *Introduction to Newton's 'Principia'* (Cambridge, Mass., 1971) p. 162.

pointing away from the sun, consist, he believed, of a vapor or finely divided reflecting matter emitted from the comet's head when it is strongly heated by its proximity to the sun.[2]

Firmly convinced that all things in nature reflect the divine purpose, and play some definite role in the plan of the universe, Newton's fertile imagination played with the possible reasons for the striking phenomenon of the tails of comets. In a long passage in the *Principia*, retained in subsequent editions, Newton described how the vapor forming the tails of a comet may be dissipated and spread throughout the heavens and by the attraction of gravity reach the earth and the other planets. Mingled with the atmosphere, it may prove to be the key to all vegetative life. To conserve the oceans, source of the rain that waters the earth and nourishes all vegetation, "comets seem to be required." For, he adds,

> from their exhalations and vapors condensed, the wastes of the planetary fluids spent upon vegetation and putrefaction, and converted into dry earth, may be continually supplied and made up; for all vegetables entirely derive their growth from fluids,[3] and afterwards, in great measure, are turned into dry earth by putrefaction . . . and hence it is that the bulk of the solid earth is continually increased; and the fluids, if they are not supplied from without, must be in a continual decrease, and quite fail at last.[4]

This long excursus ends by alluding to a view prevalent in the seventeenth century that in the air there is a special substance, a nitro-aerial spirit, necessary to sustain combustion, respiration, and the life of living creatures:

> I suspect, moreover, that it is chiefly from the comets that spirit comes, which is indeed the smallest but the most subtle and useful

---

[2]*Principia* (1687), p. 499, and Motte-Cajori, p. 522.

[3]Doubtless a reference to the well-known experiments of Van Helmont and Robert Boyle allegedly proving that water was the sole nutrient of plants.

[4]Motte-Cajori, pp. 529–530. Cited at length by Douglas McKie, "Some Early Work on Combustion, Respiration and Calcination," *Ambix*, 1 (1938), 162–163. For the Latin original see *Principia* (1687), p. 506.

part of our air, and so much required to sustain the life of all things about us.[5]

Well before he began to grapple with the dynamics of planetary motion, that is, before Edmond Halley's famous visit to Cambridge in 1684, which impelled Newton to compose the *Principia*, Newton had been studying cometary motion from a strictly descriptive and astronomical point of view. From his correspondence we learn that he was thus engaged from 1681 to 1683, when his interest, like that of his contemporaries, had been aroused by the appearance in November and December 1680 of a brilliant comet. Not long after he attempted a theoretical solution, for he wrote Halley in June 1686 that Book III of the *Principia* was written but did not yet include a theory of comets; in the autumn of 1685, he explained, "I spent two months in calculations to no purpose for want of a good method."[6]

Newton's first observations of a comet were made much earlier, and indeed are recorded in that early student notebook which he used as a commonplace book under the title of *Quaestiones quaedam philosophicae*. For the interest aroused by this comet of 1664, and the excitement it caused in the newly founded Royal Society of London, we can do little better than turn to the volumes of the *Correspondence* of Henry Oldenburg for the eventful years 1663 to 1667 as made available to us by A. Rupert Hall and Marie Boas Hall.[7] With these two volumes of the letters from and to Oldenburg, we enter upon one of the most active and fascinating periods in the early history of the Royal Society of London.

The years 1663–1667 were testing ones for the Society, years in which Oldenburg as junior Secretary—the senior Secretary, John Wilkins, played only a nominal role—turned his post into the most significant of that body. The President, Lord Brouncker, who held

[5]The Latin reads: Porrò suspicor spiritum illum, qui aeris nostri pars minima est sed subtilissima & optima, & ad rerum omnium vitam requiritur, ex Cometis praecipue venire. *Principia* (1687), p. 506.

[6]Cited by John Herivel, *The Background to Newton's* Principia (Oxford, 1965), p. 99.

[7]Oldenburg *Correspondence*, II, 1663–1665; III, 1665–1667 (1966).

his office from the foundation of the Society until 1667, was more active than the majority of those who succeeded him, yet he can hardly be said to have influenced as much as Oldenburg the course of the Society's progress. If Robert Boyle and Robert Hooke were the scientific luminaries of the early Royal Society, Henry Oldenburg was its man for all seasons. The chief duty of the secretaries, besides supervising the clerks who entered matters into the Society's records, and having custody of the Statutes, Journal Books and other documents, was to carry on domestic and foreign correspondence in the name of the Society. Such official communications, together with letters of a more personal sort, make up the bulk of these two substantial volumes.

By far the greatest number of these early letters were exchanged between Oldenburg and his friend and patron, Robert Boyle. Several of Boyle's letters make their appearance here for the first time; interesting letters, but characteristically crabbed and devoid of stylistic elegance. The reader notes, too, in the first volume the effort of Oldenburg to extend and enrich his scientific contacts with men of the European Continent: with earlier correspondents like Spinoza, with members of the Montmor Academy in Paris, with Christiaan Huygens, with the great astronomer, Johann Hevelius, in Danzig. Oldenburg describes for his European correspondents the activities of the Society: Hooke's attempts at devising a machine to grind telescopic lenses and the publication of his *Micrographia*; the books of Robert Boyle that were emerging from the press in these years; the plan for Sprat's well-known *History* of the Society. Conversely, his English correspondents received news from the Continent, were informed of Campani's new telescopic lenses, of the activities of Huygens, the publication of Grimaldi's great *De Lumine,* and G. D. Cassini's discovery that the planet Jupiter rotates on its axis. When Oldenburg wrote Robert Moray about the appearance of Athanasius Kircher's *Mundus subterraneus,* Moray replied (Letter 438) that whatever other men thought of Kircher, "I assure you I am one of those that think the Commonwealth of learning is much beholding to him though there wants not chaff in his heape of stuff . . . yet there is wheat to be found almost everywhere in them."

Of particular interest to Oldenburg and his colleagues were the steps being taken to create a similar scientific organization in France. Oldenburg describes to Boyle the meeting of the Royal Society on 10 June 1663 ("no ordinary meeting") attended by two Frenchmen associated with the Montmor Academy, Balthazar de Monconys and Samuel de Sorbière (both of whom promptly wrote travel books about their experiences), and two gentlemen from Holland, Christiaan Huygens and his father, Constantyn. Several letters are devoted to the unfortunate circumstances that attended the publication of Sorbière's book. A short time later Oldenburg reports to Boyle that Louis XIV is bestowing pensions upon several learned men, "but most poets and Romancers, except Huygenius, and Hevelius, and La Chambre; having neglected Roberval, Fermat, Frenicle, Rohaut, Ozou, and such like, qui colunt Musas severiores." A letter of Spinoza to Oldenburg next broke the news that Huygens was planning a "transmigration into France," a fact that Oldenburg passed on to Boyle. And finally, in November of 1665, Moray wrote the Secretary (Letter 462) that "Colbert intends to sett up a Society lyke ours & make Hugens Director of the designe."

Some light is thrown, here and there, on obscure figures. For example, there are a number of references to Thomas Streete, author of the *Astronomia Carolina* (1661), which the editors could well have pointed to as the textbook from which Newton acquired his first recorded notions of planetary astronomy. That they did not is cause for surprise, since Newton's notes from Streete occur in that student notebook which one of the editors of these volumes was the first to bring to light.[8]

We learn something, too, about Giuseppe Borri, the Italian alchemist and medical quack whom Oldenburg had earlier encountered in Amsterdam. Perhaps he might have been identified with

[8]Cambridge University Library, MS. Add. 3975. See A. R. Hall, "Sir Isaac Newton's Note-Book, 1661–65," *Cambridge Historical Journal*, 9 (1948), 239–250. For the entries "Out of Streete," see fols. 27v–30v.

the mysterious "Bory," a man possessed of secrets "of great worth both as to medicine & profit" and who "usually goes clothed in green" of whom Newton wrote in his famous early letter to Francis Aston.[9]

Important letters date from the summer and autumn of 1665 when London was beset by the great plague. On its account the Royal Society went into recess late in June, not to reconvene until the following February, and its members scattered to the country. All, that is, except Henry Oldenburg, who remained steadfastly at his post, worrying about how the next number of the *Philosophical Transactions* could be printed, carrying on his foreign correspondence and writing to those who, like Boyle and Moray, had retreated to safer ground. Late in August he informed Robert Hooke that "the sickness grows hotter here," adding quickly that men of naturally healthy constitutions who live "orderly and comfortably" and not "closely and nastily" seem not to contract the disease. And to Boyle Oldenburg sends a sample of Borri's presumed antidote to the plague, though for fear of upsetting his careful regimen he says he barely tasted it; Boyle's reply is notably cautious and skeptical. Owing to the plague, something resembling the old scientific group at Oxford was temporarily reconstituted. Moray writes of "our Society" having met at Boyle's house to see experiments on color performed by the host, and he lists among those present William Petty, John Wallis, John Graunt, and Sir Paul Neile. And Boyle confirms that "there being now at Oxford no inconsiderable number of ye Royall Society," he has proposed regular meetings on Wednesdays, where on one such occasion, over a dish of fruit, "we had a great deale of pleasing Discourse" and, somewhat ironically, drank to Oldenburg's health. In reply, Oldenburg expressed pleasure at this gathering of the faithful and hoped they might serve to convert "ye Oxonians" to the kind of solid knowledge the Royal Society pursued, and away from "Scholasticall contentions."

There is, however, one subject, running as a persistent thread through many of the letters in both these volumes, that the editors

[9] Newton *Correspondence*, I (1959), 11.

have not thrown into full relief, and understandably, because its significance, indeed, has quite generally eluded the historian of science. This is the excitement caused by the appearance, early in December 1664, of a brilliant comet, the most astonishing to appear in the heavens since the first years of the seventeenth century.[10] To appreciate the effect it produced, it is sufficient to point out that no comet is recorded in the period from 1618 to 1652–1653, and that Descartes, who devoted long pages in his *Principia* to a theory of comets, had probably never really observed one. The comet of 1652–1653, like that of 1661–1662, caused little public stir—both made only brief appearances; they were the concern chiefly of dedicated astronomers like Hevelius, who was led by these two events to undertake his *Cometographia,* a massive work (he described it to Oldenburg, not inaccurately, as *opus satis prolixum*) that did not appear until 1668.

The comet of 1664, by contrast, created a sensation: for three weeks after perihelion it remained tolerably close to the earth, and of remarkable brilliance. Patient astronomers could observe it as late as March. Now this comet chose to arrive at a time when precision astronomy had made remarkable advances, and when scientists of England and Europe were beginning to enjoy that organized and regular communication which the newly founded Academies made possible and which Oldenburg was busy fostering. It made a profound impression upon the scientific community; it was observed in every part of Europe, and in America as well, by astronomers of varying reliability, and it led to a substantial literature of books and pamphlets. The list of those who followed the comet, who attempted to determine its real path or who, because of it, speculated about the nature of comets in general, included many of the outstanding names of the time: Adrien Auzout and Pierre Petit in France; G. D. Cassini in Rome; Huygens in Holland; Borelli at Pisa and, of course, Hevelius in Danzig. Heve-

[10]The comet was first observed, though not in England, on 17 November 1664; it reached perihelion on 4 December. See J. Holetschek, "Untersuchungen über die Grösse und Helligkeit der Kometen, I. Die Kometen bis zum Jahre 1760," *Akademie der Wissenschaften, Wien; Mathematisch-naturwissenschaftliche Klasse. Denkschriften,* 43 (1896), 462.

lius in fact deferred the completion of his *Cometographia* until he could observe and report on this new visitor, and devoted to it his *Prodromus cometicus* of 1665. In England it became the center of attention; and for two years or more this comet, and a second even brighter one that appeared briefly in April 1665, were the subjects of continual discussion and speculation. These volumes of the *Correspondence of Oldenburg* have innumerable references to them, references that supplement the information scattered through Birch's *History of the Royal Society*. Robert Hooke and, for a time, Christopher Wren were especially charged by the Society with observing the comet of 1664 and correlating their observations with those that came in from other sources.[11] Hooke, as we shall see, soon lectured on it at Gresham College. Not surprisingly, the comet bulks large in the very first number of the *Philosophical Transactions*.[12]

A point to be noted is that the comet of 1664 led to the first concerted attempts since the time of Kepler not only to determine the parallax of a comet and its distance from the earth—in itself no easy matter—but to work out its true path. No real understanding of the later achievement of Newton and Halley in treating the comets of 1680 and 1682 as members of the solar family and computing their paths can be gained without an appreciation of the pioneer studies of this comet of 1664. Tycho Brahe—assuming, to be sure, that the earth is at rest—had assigned to comets a true path in a circle about the sun, beyond the orbit of Venus. But the commonly accepted view of those willing to concede the earth's motion—and this included most Englishmen—was that of Kepler, who believed comets to be adventitious visitors to our solar system, making a single appearance only to disappear forever. The true path, Kepler argued, was a straight line; any observed curvature of the

[11]As the Halls explain (Oldenburg *Correspondence*, II, 429, n. 4), Wren was relieved of this assignment early in 1665 and "commissioned by Charles II to prepare a scheme for rebuilding St. Paul's Cathedral (not yet destroyed by the Fire)." Hooke's colleagues were annoyed by his dilatoriness. See letters Nos. 418 and 454 of Moray to Oldenburg. Though Hooke presented his results to the Royal Society on 8 August 1666, they were not published until 1678.

[12]*Phil. Trans.* No. 1 (6 March 1664/65), 3–8; No. 2 (3 April 1665), 17–20; No. 3 (8 May 1665), 36–40; and No. 6 (6 November 1665), 104–108.

path was attributable to the earth's annual motion. This view we find accepted, for example, by Thomas Streete in his *Astronomia Carolina,* and by John Wallis in a letter to Oldenburg written not long after the appearance of the comet.[13]

But careful observations of the comet of 1664 led to two significant conclusions. The first was that the true paths of comets, not merely the apparent paths, were curvilinear, although not necessarily circular; indeed both Borelli and Hevelius suggested that the visible portion of the comet's orbit might be a conic, perhaps a segment of an ellipse. The second inference was that comets might follow closed paths with periods short enough to allow, in some cases, their return within the astronomical record. Pierre Petit, in his *Dissertation sur la nature des comètes* (1665), was apparently the first to suggest that a comet could return after a given lapse of time.[14] The comet of 1664, he suggested, might be the same one that had appeared 46 years earlier in 1618. From the diary of Samuel Pepys we learn that the same thought occurred to Robert Hooke, who made the suggestion in "a very curious lecture about the late Comet" delivered at Gresham College on 1 March 1664/65. This, said Pepys, "is a very new opinion."[15]

The comet would seem to have exerted an even more fundamental historical role. If I am correct, it made a real contribution to the acceptance of the Copernican hypothesis. Pierre Petit, though doubtful about the earth's annual revolution, accepted the diurnal rotation and—like Auzout—argued that the Church had not pronounced definitely on the question of the earth's motion.

[13]Aware that conjectures about cometary paths "be now stirring," Wallis in this important letter describes the cometary hypothesis found among the papers of Jeremiah Horrox. Horrox imagined that comets originate in the sun and return to it in a curved path "by an Ellipticall figure or near it." Letter No. 366, Wallis to Oldenburg, 21 January 1664/65. To be sure, Thomas Hariot, long before, had proposed elliptical cometary paths *about* the sun.

[14]Petit's book has been unduly neglected. Andrew D. White called it a "vehement attack on superstition, addressed to the young Louis XIV" (*History of the Warfare of Science with Theology* [New York, 1897], I, 198). G. Atkinson in a valuable paper treats Petit as a precursor of Bayle and Fontenelle, and has gathered useful information on the comet of 1664. See his "Précurseurs de Bayle et de Fontenelle," *Revue de littérature comparée,* 25 (1951), 12–42.

[15]*The Diary of Samuel Pepys,* ed. Henry B. Wheatley, IV, 341. Hooke's lecture (Pepys describes it as his *second* one on the comet) was an afternoon public lecture given in his capacity as Professor at Gresham College.

Auzout, who had the temerity to publish an ephemeris of the comet on the basis of four or five observations, went so far as to suggest
that its study might "serve to decide the grand Question of the
Motion of the Earth."[16] And Hevelius in his *Prodromus* flatly stated, as
Oldenburg rendered his words in the abstract he published, "that
without the *annual Motion* of the *Earth*, no rational Account can be
given of any comet, but that all is involved with perplexities, and
deform'd by absurdities."[17]

The historical role of this comet has yet to be explored in detail.
In France it awakened the curiosity of the Court; at Royal com-
mand, a conference was held early in January at the Jesuit Collège
de Clermont, attended by great noblemen and prelates, where
scholars set forth conflicting views about the nature of comets.[18]
Pierre Petit dedicated his *Dissertation sur la nature des comètes* to Louis
XIV, at whose behest, in fact, it was composed, taking the oppor-
tunity, as did Auzout, to urge the establishment of an Observatory
and an officially sponsored Royal Academy. Besides this, the study
of the comet surely gave a great impetus to astronomical investigation, and perhaps it contributed to the perfecting of the famous
filar micrometer, described by Auzout in a letter to Oldenburg
written late in December 1666.[19]

Lastly there is Newton, whose commonplace book records his
observations of the comet from 17 December to 23 January 1664/
65. These observations, I suggest, may well have been responsible
for arousing Newton's earliest interest in technical astronomy. His
notes from Streete's *Astronomia Carolina*, and a clear-cut reference
to Thomas Salusbury's translation of Galileo's *Dialogo*, seem to have
been set down shortly after these observations of the comet. His
astronomical jottings must have been written late in 1665 or early
in 1666 (N.S.).[20]

[16]*Phil. Trans.*, No. 1 (6 March 1665), 6.

[17]*Phil. Trans.*, No. 6 (6 November 1664–65), 105.

[18]See Atkinson, pp. 27–28. The meeting was described at some length in the
Journal des Sçavans.

[19]Letter 589 and *Phil. Trans.*, No. 21 (21 January 1666/67), 373–375.

[20]They follow immediately after a reference (MS. Add. 3975, fol. 20v) to Auzout's estimates of the apparent diameters of certain stars. Auzout's name is barely decipherable, but the figures Newton ascribes to him are those that appear in

Not long afterward, as everyone knows, probably in the summer of 1666, Newton made his now legendary inquiry into the force of gravitation. Was there perhaps a connection? It is at least possible. Why, indeed, should Newton have asked himself, as he did in the garden at Woolsthorpe, whether the same force that draws a body to the earth holds the moon in its orbit? Descartes had provided a plausible explanation for both effects through the vortical motion of his "second matter." Newton was thoroughly familiar with Descartes's speculations, as his notebook makes quite clear. What led him, then, to question the Cartesian explanation? Doubtless there were many factors, but something striking must have led him to question for the first time, and eventually reject, Descartes's vortices. The comet of 1664, for a special reason, could well have played this role in Newton's thought: it was a *retrograde comet,* one that moved East to West, *opposite* the direction of the planets. Clearly—as a mind as keen as Newton's would have perceived—if the comet penetrated the solar system, as it obviously did, its motion was totally uninfluenced by the vortical stream that Descartes invoked to explain the planetary revolutions. In view of the stress Newton was later to place on comets, both in the *Principia* and the *Opticks,* as evidence against the Cartesian vortices,[21] it may very well have been the comet of 1664 that shook him free from Cartesian explanations, led him to explore the mysteries of celestial motion and invoke the idea, already being mooted but widely discredited, of an attractive force operating between the bodies of the solar system.

---

an abstract of a paper of Auzout published in *Phil. Trans.,* No. 4 (5 June 1665), 57–63.

[21]See, for example, *Principia* (1687), p. 480 (Bk. III, Prop. XXXIX, Lemma IV, Corol. 3), where Newton emphasizes the retrograde motion of some comets: "Hinc etiam manifestum est, quod coeli resistentia destituuntur. Nam Cometae vias obliquas & nonnunquam cursui Planetarum contrarias secutui, moventur omnifariam liberrimè, & motus suos etiam contra cursum Planetarum diutissimè conservant." Motte-Cajori (p. 497) renders the passage thus: "Hence also it is evident that the celestial spaces are void of resistance; for though the comets are carried in oblique paths, and sometimes contrary to the course of the planets, yet they move every way with the greatest freedom, and preserve their motions for an exceeding long time, even when contrary to the course of the planets." In later editions of the *Principia* the section on comets is greatly expanded and Newton not only invokes Halley's results, but devotes considerable space to the observations on the comet of 1664 made by Hevelius, Auzout, Pierre Petit, and Hooke, all of whom are named.

If the influence on Newton of the comet of 1664 is conjectural and admittedly rather bold, the evidence is somewhat clearer in the case of Robert Hooke, for it was precisely during the period in which he was charged by the Royal Society to observe the comet, correlate his observations with those of others, and come forward with a "hypothesis" to explain its motions, that Hooke turned to the problem of gravity. On 21 March 1665/66 Hooke read to the Society his first speculations.[22] He soon concluded that all celestial bodies move in closed paths about the sun (or some parent planet) because their tangential motion is coupled with the action of a "supervening attractive principle." This he tried to demonstrate by his well-known, and often criticized, experiments with a circular pendulum. He reported his results at the Society's meeting of 23 May 1666, announcing on this occasion that by his hypothesis "the phenomena of the comets as well as of the planets may be solved."[23] Like Sir Christopher Wren, Hooke pursued his inquiry about gravitation to the point of hitting upon (though he, like Wren, was far from proving it) the law of attraction according to the inverse square of the distance. It is indeed remarkably coincidental that two of the three men—Hooke, Wren, and Halley—whom Newton was persuaded to acknowledge in the *Principia* as independent discoverers of the law of gravity, were the men especially charged with the study of the comet of 1664.

Whether these conjectures have any validity or not, there can be little doubt of the important historical reverberations of the comet of 1664, and it deserves to be rescued from virtual oblivion. We have, I think, been bemused by the even greater public excitement aroused by Bayle's comet of 1680 and by the significant results that Halley derived from that of 1682 (Halley's comet), and so we have left the earlier one in the lurch. Yet the intense activity that attended its appearance paved the way for the successful elucidation of the two later, and more widely discussed, seventeenth-century comets.

[22]Birch, *History of the Royal Society*, II, 69–72. Hooke sought to determine whether the force of gravity varies with the distances from the center of the earth.
[23]Ibid., p. 92.

CHAPTER 3

Newton on the Continent

The Early Reception of His Physical Thought

B esides the technical study of Newton's achievements in mathematics, optics, and dynamics, there is a phase of Newtonian
scholarship which has attracted renewed interest and which we
may call the "influence," the "reception," or the "legacy" of Newton. This is ambiguous, of course, for there are at least two ways in
which the subject can be viewed: we can consider Newton's reception by his learned contemporaries or his influence upon his scientific successors (by any definition of what the history of science is
about, surely this is of major importance); or, on the other hand,
we can deal with the influence he exerted through his natural
philosophy and his advocacy of the experimental approach—that
is, those aspects of his achievement comprehensible to the intelli-
gent layman—upon the *Weltanschauung* of his age or later times.
Clearly, this is a legitimate part of cultural or intellectual history, as
scholars like Daniel Mornet and Preserved Smith (to name only
two) long ago perceived.[1] And within our ranks of historians of
science, Hélène Metzger, with her book on what she called the
Newtonian "commentators," was surely an outstanding pioneer.[2]

From *History of Science*, 17 (1979). Originally entitled "Some Areas for Further
Newtonian Studies," this paper was presented at a Colloquium held at Churchill
College, Cambridge, in August 1977 and devoted to this subject. The problems
raised here were first gone over in my graduate seminar at Cornell in 1962, and I
am indebted to the stimulus of my students.

[1]Daniel Mornet, *Les sciences de la nature en France au XVIIIe siècle* (Paris, 1911);
Preserved Smith, *A History of Modern Culture*, II (New York, 1934), esp. chaps. 2–4.

[2]Hélène Metzger, *Attraction universelle et religion naturelle chez quelques commentateurs
anglais de Newton* (Paris, 1938).

I hasten to add that this second aspect of Newton's "legacy" has little or nothing to do with the so-called "externalist" approach to the history of science. It projects outwardly from science, or a branch of science, to observe its reception by society, instead of pointing inwardly to seek social, economic, and intellectual influences upon a scientist's work, influences which I happen to believe we cannot, in many cases, safely ignore. The current externalist–internalist debate—like many "either-or" disputes—seems utterly contrived and fatuous.[3]

Mme Metzger's book is a classic; yet she did not seriously raise, or try to answer, the question why the Boyle lecturers, the popularizers like John Keill and Henry Pemberton, or mavericks like John Toland, felt it important to interpret Newton's new system of the world for their contemporaries. The problem has recently been tackled by Margaret Jacob in various articles and in her rather audacious book, where she has argued that the popular exposition of Newton's world view by these "commentators" served ideological, religious, and sociopolitical ends.[4]

As to the reception of Newton in France, we are left—despite the books of Pierre Brunet, Ira Wade, and others—with considerable murkiness and oversimplification; and for the early period, say from 1672 to 1699, we have only tantalizing allusions, and scattered nuggets, in books written with other purposes in mind. In this paper I should like to single out two areas, two aspects of Newton's reception in France, that seem to call for more systematic investigation, and in the second area, for some reassessment.

The first has to do with Newton's reputation on the Continent before 1699, the year in which he was made a foreign associate of the Royal Academy of Sciences in Paris. The second, which I shall treat in some detail, is the reassessment of the conventional pic-

[3]The distinction between "internal" and "external" influences upon science is older than the present-day methodologists of the history of science may realize. The distinction—obvious enough in itself—and even the precise terminology appeared as early as 1948 in a paper presented by Jean Pelseneer to the Comité Belge d'Histoire des Sciences, and published as "Les influences dans l'histoire des sciences," in the *Archives internationales d'histoire des sciences*, 1 (1947–48), 347–353.

[4]Margaret C. Jacob, *The Newtonians and the English Revolution, 1689–1720* (Ithaca, N.Y., 1976).

ture—of Cartesians arrayed against Newtonians—which Pierre Brunet sets forth in his study of the introduction of Newton's physical theories in France before 1738, that is, before Maupertuis, Clairaut, and Voltaire hoisted the banner of Newtonian physics in France.[5]

When did scientists on the Continent first hear of Newton? By what stages did his reputation grow until, for his recognized accomplishments, he was honored by the Academy of Sciences in 1699? We can, I think, pass over the flattering reference to him by name, as early as 1669, in the preface of Isaac Barrow's lectures on optics, or the appearance of the edition of Varenius's *Geographia generalis* Newton published in Cambridge in 1672. Neither could have aroused much interest abroad in this obscure, if obviously very gifted, young Englishman.

The chief events that brought Newton's name before the European scientific public were doubtless: (1) the publication in February 1672 of his now-famous first letter on light and color;[6] (2) his invention of the reflecting telescope, which brought about his election as F.R.S. in January 1672; and (3) Henry Oldenburg's orchestrated propaganda on Newton's behalf.

It was Newton's stubby little telescope, which promised to eliminate the chromatic aberration—the colored fringes—invariably encountered in the refracting telescopes of that day, that excited Christiaan Huygens (and others) when Oldenburg wrote him about Newton's discoveries.[7] In the new theory of the origin of

[5]Pierre Brunet, *L'introduction des théories de Newton en France au XVIIIe siècle*, I: *Avant 1738* (Paris, 1931). This one volume carrying the story to 1738 was all that was published.

[6]*Phil. Trans.*, 6 (1671–72), No. 80, 3075–3087. Newton's "Accompt of a New Catadioptrical Telescope" was published later in the same volume, 4004–4010. These optical papers may be conveniently consulted in *Newton's Papers*, pp. 47–67.

[7]Oldenburg first informed Huygens of Newton's new kind of telescope by letters of 1 and 15 January 1671/72. See *Oeuvres de Huygens*, VII (1897), 124–125, and 128; also Newton *Correspondence*, I (1959), 72–76, and 81–82; and Oldenburg *Correspondence*, VIII (1971), 443–445, and 468–473. Greatly impressed, Huygens sent a description of Newton's telescope, with a letter giving his opinion of it, to Jean Gallois, editor of the *Journal des sçavans*. Gallois published both in the issue of 29 February 1672. Other French savants, among them Adrien Auzout, Jean-Baptiste Denis, and of course Cassegrain, were interested in Newton's invention.

color, Huygens evinced little interest, although Oldenburg repeatedly importuned him for his opinion.[8]

The first French savant to give serious attention to Newton's early paper on light and color was the Jesuit scientist, Father Ignace Gaston Pardies (1636–1673). This has always struck me as curious. Pardies was not a member of the Academy of Sciences, nor had he any longstanding contact with the Royal Society.[9] But he was a friend of Huygens, and a frequent participant at the meetings of the so-called Academy of the Abbé Bourdelot, an informal society, antedating the creation of the Academy.[10] After 1666, the membership of Bourdelot's group consisted chiefly of persons whom the Academy, for one reason or another (such as being, like Pardies, a Jesuit) did not admit; but members of the Academy, Huygens among them, were sometimes seen *chez* Boudelot.[11] Huygens, in any case, was aware that Pardies had a keen interest in optics, indeed was at work on a treatise that made use of a wave or pulse theory of the nature of light.[12] Perhaps Huygens, preoccupied with other matters (he had not begun to work out the theory he developed in his *Traité de la lumière* of 1690) brought Newton's

[8]In March 1672 Oldenburg sent Huygens the number of the *Transactions* containing Newton's pioneer paper on light and color, asking for his opinion of the new theory. See *Oeuvres de Huygens*, VII, 156; Newton *Correspondence*, I, 117; and Oldenburg *Correspondence*, VIII, 584–585. Huygens contented himself with replying: "Pour ce qui est de sa nouvelle Théorie des couleurs, elle me paroit fort ingenieuse, mais il faudra veoir si elle est compatible avec toutes les expéiences" (*Oeuvres de Huygens*, VII, 165). With slight variations this passage was quoted in a letter of Oldenburg to Newton (19 April 1672), in Newton *Correspondence*, I, 135.

[9]Pardies's *Discours du mouvement local* (Paris, 1670), published anonymously, had been translated by Oldenburg and published in London that same year. Pardies's first formal communication was a flattering letter to Oldenburg, dated 18 July 1671, remarking that he had just been shown parts of the *Philosophical Transactions* and learned that Oldenburg had translated his *Discours* into English (Oldenburg *Correspondence*, VIII, 143–145).

[10]The Academy of the Abbé Bourdelot has been described in Harcourt Brown, *Scientific Organisations in Seventeenth Century France (1620–1680)* (Baltimore, 1934), chap. 11.

[11]Huygens made an early reference to the Bourdelot group in a letter written from Paris on 26 April 1664 to his brother Lodewijk. See Brown, p. 233.

[12]Pardies did not live to complete his book on optics; but material from his draft was used, with full acknowledgment, by Father Pierre Ango, a fellow Jesuit, in his *L'optique divisée en trois livres* (Paris, 1682), unpaginated dedicatory preface and p. 14. If Pardies had accepted Newton's theory of color, there is no trace of it in Father Ango's book, where all the colors are explained according to the old theory of a mixing of black and white.

paper to the attention of Pardies, with the suggestion that he evaluate it. All this, of course, is conjectural, but it may explain why Pardies's unsolicited paper, a letter to Oldenburg dated 9 April 1672, was the first response any French scientist made to Newton's new theory of color.[13]

Except for one element of confusion, the story of the cool reception of Newton's challenging new theory is all too familiar, and need not be repeated here. This element deserves at least a passing mention, otherwise it would be hard to understand the lack of interest in Newton's theory of color during these early years. Pardies never claimed to have repeated successfully Newton's famous two-prism experiment, the so-called *experimentum crucis*. Yet English or American readers could readily believe that this is so, if they confine their attention to the English version of the concessive phrases of Pardies's letter of 9 July 1672. The mistranslation of these important sentences appeared in the *Philosophical Transactions Abridged* (1809), which I. Bernard Cohen chose to use in his *Isaac Newton's Papers and Letters on Natural Philosophy*.[14] This English version reads, with reference to the *experimentum crucis*, "When the experiment was performed after this manner, *everything succeeded*, and I have nothing further to desire."[15] Pardies, in fact, having finally understood the experiment, with the help of a sketch supplied by Newton, had simply written: "L'expérience ayant esté faite de cette façon je n'ay rien à dire." There is no reference to an experiment "succeeding."

At all events it was Newton's optical work, together with the reflecting telescope, that first made his name familiar in French scientific circles.[16] The experimental results he claimed, and the

---

[13]For Pardies's first letter see Newton *Correspondence* I, 130–133.

[14]*Philosophical Transactions Abridged* (London, 1809), VII, 743, and *Newton's Papers*, p. 109. The three-volume abridgement of the early *Transactions* by John Lowthorpe (London, 1705), gives Pardies's statement of concession only in Latin (I, 144).

[15]The italics in the quotation are my own.

[16]For the French original of this passage see Newton *Correspondence*, I, 205–206. The error is not attributable to Oldenburg's translating the French into Latin, the language in which it appears in the original *Transactions*, for what we read is a good rendering of the French: "Experimentum peractum cùm fuerit isto modo, nil habeo quod in eo desiderem ampliùs (*Phil. Trans.*, 7 [1672–75], 5018, reproduced in *Newton's Papers*, p. 103).

ingenuity of his theory, made it impossible to neglect his results for long. About 1679 France's leading experimental scientist, Edme Mariotte, determined to confirm or refute the Newtonian doctrine of color. He successfully repeated a number of Newton's experiments, but when he tried the *experimentum crucis* he concluded that the rays separated by the first prism did not appear to be monochromatic; on the contrary, they seemed to be further modified by the action of the second prism, yielding fringes of different colors.[17] Newton's theory, he concluded, could not be accepted. For more than a generation, this was gospel in France, and Newton's theory of color remained in disfavor.[18]

It is generally agreed, although the supporting evidence has not been fully marshalled, that Newton's reputation in these years as a meteoric mathematical genius outshone his work on light and color or even the invention of his reflecting telescope. Although he had published none of his mathematical discoveries before the appearance of his *Principia*, his repute as a mathematician of extraordinary ability was clearly established.

As historians of ideas we are happiest when we can navigate from the firm ground of one document to the next, and we are prone to forget how great a part travel, gossip, and word-of-mouth have played in the diffusion of scientific knowledge, indeed of knowledge of all sorts. We are truly fortunate when surviving letters or memoranda give us some hint of these informal exchanges.

In 1669, the self-educated London mathematician, John Collins, that "clearing house for mathematical gossip," as D. T. Whiteside has called him, learned from Isaac Barrow of Newton's *De analysi*, the earliest of Newton's mathematical papers to be circulated.[19] It

---

[17]Edme Mariotte, *De la nature des couleurs* (Paris, 1681), p. 211. This paper was reprinted in the *Oeuvres de Mariotte*, 2 vols. (Leiden, 1717), I, 227–228, and in a later edition of the *Oeuvres*, 2 vols-in-one (The Hague, 1740), consecutively paginated. The supposed refutation of Newton's experiment appears on pp. 227–228 of this edition.

[18]For new material on the later stages of the penetration of Newtonian optics into France see A. Rupert Hall, "Newton in France: A New View," *History of Science*, 13 (1975), 233–250.

[19]Derek T. Whiteside, *The Mathematical Works of Isaac Newton* (New York and London, 1964), p. xii. Whiteside reprints in facsimile John Stewart's 1745 English

contained an outline of his discoveries concerning infinite series and various applications of series expansion, but only a hint of the fluxional calculus. Collins, with his extensive contacts on the Continent, communicated Newton's results, but apparently not the precise methods used, to Slusius in Holland, to Jean Bertet and Francis Vernon in Paris, and to the venerable Giovanni Borelli in Italy. In May 1672, soon after he had learned about Newton's optical investigations, Christiaan Huygens heard from Henry Oldenburg that Newton had in hand an enlargement and corrected version of Mercator's Latin translation of Kinckhuysen's *Algebra,* a work for which Newton, it turned out, never found a publisher.[20]

It was not long before Leibniz—diplomat, philosopher, and polyhistor, and soon to emerge as Europe's most brilliant mathematician—learned about Newton. We are so accustomed to thinking of these two great men in terms of their later dispute over the invention of the calculus or of their philosophical differences set forth in the letters of the Clarke–Leibniz correspondence, that we overlook their earlier relations, or at least the occasions on which Leibniz spoke of Newton with admiration.

In 1672 Leibniz came to Paris on a diplomatic mission, chiefly designed to dangle before Louis XIV the proposal that he embark on the conquest of Egypt to satisfy the French monarch's martial ambitions and to wean him away from invading the Low Countries and Germany. The plan was offered too late—although, long after, it appealed to Napoleon Bonaparte—for the troops of the Sun King were already on the march. During this visit, however, Leibniz formed a close tie with Christiaan Huygens, met other members of the Academy of Sciences, and must have learned something about Newton's reflecting telescope and the new theory of color.[21]

---

translation of the *De analysi.* For the original Latin version, a new translation, and illuminating notes, see Whiteside, *Newton's Mathematical Papers,* II (1968), 206–247.
[20]Newton *Correspondence,* I, 155–156. For Newton's work on the Kinckhuysen *Algebra,* see Whiteside, *Newton's Mathematical Papers,* II, 277–291.
[21]Henri L. Brugmans, *Le séjour de Christian Huygens à Paris* (Paris, 1935), pp. 72–73. In August 1676 Leibniz wrote to Oldenburg: "Inventa Neutoni ejus ingenio digna sunt, quod ex Optices experimentis et Tubo Catadioptrico abunde eluxit" (Newton *Correspondence,* II (1960), 57). Oldenburg had earlier drawn Huygens's attention to Leibniz, mentioning in a letter of late March 1671 Leibniz's *Hypothesis*

Early the following year Leibniz was in London where he surely heard echoes of Newton's mathematical prowess, for he came to know various English scientists (among them Oldenburg, Robert Boyle, and the mathematician John Pell) and attended a meeting of the Royal Society.[22] Perhaps he met John Collins who, despite a lack of university training, had been made F.R.S. in 1667; but he certainly did not see Newton, comfortably immured in Trinity College, Cambridge. Leibniz's second English visit, in 1676, was in many respects more rewarding. At Collins's urging, Newton wrote for Leibniz his *Epistola prior* (13 June) describing his generalized binomial theorem. When Leibniz asked for more information, Newton replied with his *Epistola posterior* (24 October) expounding in more detail his binomial theorem, but also giving the key to his calculus—to the general method of drawing tangents, solving problems of maxima and minima, and so on—but only by means of his famous cipher, a seemingly meaningless jumble of letters and numbers.[23] Leibniz was suitably impressed; and referring to Newton's work on series he wrote: "That remarkable man is one of the few who have advanced the frontiers of the sciences."[24]

Whatever else it may have done—and it did not produce a crop of instant Newtonians across the Channel—the publication of the *Principia* greatly enhanced Newton's stature as a mathematician. As early as 1686, European scientists heard rumors about a forthcoming book by Newton. The source was, not surprisingly, Edmond Halley, who had not only cajoled Newton into writing and publishing it, and advanced the sum for printing it, but announced it in the *Philosophical Transactions* in 1686 and by personal letter. Huygens, for his part, learned of it in June 1687 through his young

---

physica nova, his earliest study of motion, dedicated to the Royal Society. See Oldenburg *Correspondence,* VII (1977), 573–579. For this work see the article "Leibniz: Physics, Logic, Metaphysics" by Jürgen Mittelstrass and Eric J. Aiton in DSB, VIII (1973), 150–160.

[22]For Leibniz's participation in the meeting of 22 January 1672/73, see Birch, *History of the Royal Society,* III, 73. Leibniz demonstrated an early version of his calculating machine. He was elected F.R.S. on 9 April 1673.

[23]For these Latin letters to Leibniz, with English translations, see Newton *Correspondence,* II, 20–47, and 110–161.

[24]Newton *Correspondence,* III (1961), 3–5.

friend, Fatio de Duillier,[25] and wrote that he was eager to see the book.[26] His copy seems to have arrived sometime in 1688 and he read it carefully enough to comment on it that year, in what Westfall calls "a cryptic note": "Vortices destroyed by Newton. Vortices of spherical motion in their place."[27] Huygens's reference is to a theory he had defended as early as 1669 in a debate at the Academy in which Frénicle, Roberval, and others took part. While unhappy with the Cartesian vortex theory, he could not suffer Roberval's willingness to invoke an attractive power as the cause of gravity.[28] For an explanation to be intelligible Huygens argued, like the mechanical philosopher he essentially was, nothing should be invoked but matter in motion. He proposed a radical departure from the Cartesian *tourbillons*, suggesting that the circulatory motion of an aether or subtle matter took place in all planes about the earth; its tendency is everywhere centrifugal, thrusting heavy bodies toward the center.[29]

Only the advent of Newton's *Principia*, with its references, albeit cautiously phrased, to an attractive force, caused Huygens to resur-

[25]*Oeuvres de Huygens*, IX (1901), 167. Cf. I. Bernard Cohen, *Introduction to Newton's "Principia"* (Cambridge, Mass., 1971), p. 138, n. 9. For Fatio's aspirations as editor of a second edition of the *Principia*, see ibid., pp. 177–187.

[26]"Je souhaitte de voir le livre de Newton. Je veux bien qu'il ne soit pas Cartesien pourveu qu'il ne nous fasse pas de suppositions comme celle de l'attraction" (Huygens to Fatio [11 July 1687], *Oeuvres de Huygens*, IX, 190). Cf. Richard S. Westfall, *Force in Newton's Physics* (London and New York, 1971), p. 184.

[27]*Oeuvres de Huygens*, XXI, 437. See Westfall, 186. Turnbull, in Newton *Correspondence*, III, 2, n. 1, avers that Huygens "had recently" received his copy of the *Principia* from his brother Constantyn. Turnbull's reference is to a letter of Christiaan to Constantyn dated 30 December 1688 (*Oeuvres de Huygens*, IX, 304–305) which merely tells us that Huygens had read the book before that date. But a letter of Constantyn to Christiaan, dated from Loo in Western Flanders on 13 October 1687, includes the sentence: "Dr. Stanley est allé en Angleterre et me portera encore des livres curieux. Il ne revient que vers le temps que nous irons a la Haye c'est a dire dans un mois d'icy" (*Oeuvres de Huygens*, IX, 234). This supports the notion that Constantyn was the intermediary, but suggests that Christiaan may have received his copy of the *Principia* either late in 1687 or early in 1688. William Stanley (1647–1731), Dean of St. Asaph, was chaplain to the future Queen Mary and after the accession of William III was made clerk of the closet.

[28]For Roberval's gravitational theory, similar to that advanced by Copernicus and Galileo, see "Un débat à l'Académie des sciences sur la pesanteur," in Léon Auger, *Gilles Personne de Roberval (1602–1675)* (Paris, 1962), esp. p. 179. See also Westfall, pp. 184–186.

[29]Westfall, p. 187.

rect his early theory. Evidently he talked the matter over with Fatio de Duillier, for in July of 1688 Fatio was in England and described to the Royal Society the theory that Huygens had advanced to explain gravity. He promised "with Mr. Huygens's leave" to provide the Society with a copy "thereof in writing."[30] Whether this was done I do not know. In any case Huygens published his "Discours de la cause de la pesanteur" in 1690 as an appendage to his *Traité de la lumière*. Since the body of this little treatise, the "Discourse," had been written before the appearance of Newton's *Principia*, Huygens concluded with an "Addition" making mention of some of its contributions. It was impossible, he wrote, to withhold assent from the mathematical demonstration of Kepler's laws. Newton must be correct that gravity acts throughout the solar system and that it decreases in strength in proportion to the square of the distance. But the idea of an attractive force was unacceptable; gravity must be explained in some manner by motion.[31]

In 1688 reviews in European journals began to appear: in the *Bibliothèque universelle et historique*, in the Leipzig *Acta eruditorum*, and a well-known one in the *Journal des sçavans*.[32] All attempted a summary of this complex book. But the reviewer in the *Journal des sçavans* judged the *Principia* to be the work of a mathematician (*un géomètre*) rather than that of a natural philosopher (*un physicien*). Its abstract, mathematical character was that of a work in mechanics

[30]Royal Society Journal Book for 18 January 1687/88 and 4 July 1688. Fatio had been elected F.R.S. in 1687. In May 1688 he described at a meeting of the Society the pendulum clock that Huygens had devised, and which had been "sent to the Cape of Good Hope by a person skilled in Astronomy, with design to trie what might be done in the matter of Longitude by that method of clocks" (Journal Book, 9 May 1688). When Huygens met Newton for the first time in 1689 it was on a trip to England in the company of Fatio. For the meeting at Gresham College on 12 June 1689, where Huygens "gave an account" of his forthcoming "Treatise concerning the Cause of Gravity" and had an exchange with Newton about the double refraction of Iceland spar, see Royal Society Journal Book, 12 June 1689, cited by Turnbull, Newton *Correspondence*, III, 31, n. 1.

[31]*Traité de la lumiere . . . par C.H.D.Z. Avec un discours de la cause de la pesanteur* (Leiden, 1690). The *Traité*, the *Discours*, and the Newtonian "Addition" are consecutively paginated. The *Discours* is reprinted separately in *Oeuvres de Huygens*, XXI, 451–499.

[32]Edmond Halley's laudatory review, an unabashed and rhetorical bit of promotional material, can hardly have persuaded any Continental critic of Newton. See *Phil. Trans.*, 16, No. 186 (1687), 291–297.

(in the seventeenth century a recognized branch of the so-called mixed mathematics) rather than of a work of physics. To create a physics as perfect as his mechanics, so said the reviewer, the author must substitute real motions for those he has imagined.[33]

Leibniz first read a summary of the *Principia* during a diplomatic mission to Italy in 1688, when a friend gave him some recent monthly issues of the *Acta eruditorum*. In the June issue he read "eagerly and with much enjoyment" (*avide et magna cum delectatione legi*) an account of the celebrated Isaac Newton's Mathematical Principles of Nature.[34] The book itself, which had been given Fatio de Duillier to be sent to Leibniz, reached him in Rome, where he arrived on 14 April 1689.[35]

After reading the summary of the *Principia* in the *Acta*, Leibniz—like Huygens—was inspired to put down his own views. His treatise attempting to explain the motions of the heavenly bodies (the *Tentamen de motuum coelestium causis*) was dispatched to the *Acta* from Vienna and published in the issue for February 1689.[36]

The *Principia* was not as ignored on the European continent as is sometimes believed, yet Huygens and Leibniz were the only competent, sophisticated, and persistent critics of Newton's theory of celestial motions. They studied each other's attempts to devise a theory more compatible with their adherence to the mechanical philosophy, yet not at odds with Newton's manifest discoveries. From 1690 until Huygens's death in 1695 their points of agree-

---

[33]*Journal des sçavans*, 2 August 1688 (Amsterdam, 1689), 237–238. A similar view was set forth by Malebranche who wrote in 1707: "Quoique Mr. Newton ne soit point physicien, son livre [the *Optice*] est tres curieux et tres utile a ceux qui ont de bons principes de physique, il est d'ailleurs excellent geometre . . ." (*Oeuvres complètes de Malebranche* [Bibliothèque des textes philosophiques: Directeur, Henri Gouhier], XIX [1961], 771–772). This edition will be the one cited henceforth, except in n. 47.

[34]Leibniz to Mencke, in Newton *Correspondence*, III, 3–4.

[35]In October 1690 Leibniz wrote to Huygens remarking on the "quantité de belles choses" the book contained (Newton *Correspondence*, III, 80). Leibniz's own copy of the first edition of the *Principia* was discovered in 1969 by E. A. Fellmann of Basel. Leibniz's marginal annotations have been reproduced in facsimile, together with transcriptions of the marginalia and a commentary, in *Marginalia in Newtoni Principia Mathematica*, ed. E. A. Fellmann (Paris, 1973).

[36]E. J. Aiton, *The Vortex Theory of Planetary Motion* (London and New York, 1972), p. 127.

ment and disagreement, with each other and with Newton, frequently arose in their correspondence.[37] Both men were convinced that the *tourbillons* of Descartes had to be abandoned if Kepler's empirical laws were to be explained by a "deferent matter" carrying the planets around.[38] Neither accepted Newton's use of attractive forces, or was at all certain what Newton meant by his use of the word "attraction." On certain fundamental matters the two men differed. In April of 1692 Leibniz wrote: "In rereading your explanation of gravity recently, I noticed that you are in favor of a vacuum and of atoms. . . . I do not see the necessity which compels you to return to such extraordinary entities."[39]

It should be noted that neither man attacked Newton in print, despite their differences with him. Both held him in high regard. Leibniz's position was not greatly different from the views set forth in the addition Huygens made to his "Discourse." It is true, Leibniz wrote Huygens in September 1689, that according to Newton's explanation planets "move as if there were only one motion of trajection or of proper direction, combined with gravity," yet they also move "as if they were carried along smoothly by a matter whose circulation is harmonious."[40] And he adds that he cannot abandon his deferent matter because he can find no other explanation for the fact (true as far as astronomical knowledge went in the seventeenth century) "that all the planets move somewhat in the same direction and in a single region." In a letter to Newton, written in March 1693, Leibniz is generous, and I think sincere, in his praise, yet candid as to the nature of his disagreement:

[37]Early in 1690 Leibniz received a copy of Huygens's *Traité de la lumière*, containing the Dutch scientist's *Discours de la cause de la pesanteur*. In an accompanying letter, Huygens asked Leibniz if he had modified his planetary theory after reading Newton's *Principia*, proof incidentally that Huygens had already digested the "Tentamen."

[38]For Huygens's "spherical vortex" see Westfall, p. 187. Leibniz's "harmonic circulation" of a deferent aether is described by Aiton, *Vortex Theory*, pp. 125–151, and by Westfall, *Force*, pp. 303–310.

[39]Leibniz, *Philosophical Papers and Letters*, ed. L. E. Loemker, 2 vols. (Chicago, 1956), II, 679. Huygens saw no incompatibility between his aether and the concepts of atoms and the void. He conceived of his aether as rare because each particle is porous, its component subparticles being separated by many empty spaces.

[40]Leibniz, *Philosophical Papers and Letters*, II, 681.

How great I think the debt owed to you, by our knowledge of
mathematics and of all nature, I have acknowledged in public also
when occasion offered. . . . You have made the astonishing discovery
that Kepler's ellipses result simply from the conception of attraction
or gravitation and passage [*trajectio*] in a planet. And yet I would
incline to believe that all these are caused or regulated by the motion
of a fluid medium, on the analogy of gravity and magnetism as we
know it here. Yet this solution would not detract from the value and
truth of your discovery.[41]

Leibniz's last sentence is especially interesting, for it echoes the
widely held position that Newton's brilliant explanation is a mathematical "hypothesis" that saves the phenomena, but does not provide a valid "physical" account.

Whatever their reluctance to follow Newton into the mysterious
realm of an attractionist dynamics, the readers of the *Principia*
were provided with the best illustration of his mathematical brilliance, with tantalizing glimpses of his new calculus, veiled though it
was by a geometrical, rather than an analytical, presentation. What
both Huygens and Leibniz saw in the *Principia* aroused their curiosity and their interest in the rumor that the expected Latin version of
John Wallis's *Algebra* was to contain something by Newton himself
about his new methods. Both Leibniz and the Marquis de l'Hospital
(of whom more shortly) asked Huygens, as soon as a copy should
reach him, to transcribe the Newtonian passages for them. When
Leibniz finally received his copy of the extract, he thanked
Huygens, but expressed his disappointment; much that he found
there, he wrote, was already familiar to him.[42]

I should like to turn my attention to the second area of this
proto-investigation: the need for a reassessment of the so-called
Cartesian–Newtonian debates in France before 1738. I doubt that
we can accept, without modification, the highly polarized—indeed

[41]Newton *Correspondence*, III, 257–258. "Trajection" or "projection" would be
preferable translations of *trajectio*.
[42]For this correspondence, see *Oeuvres de Huygens*, IX, letters nos. 2777, 2785,
2815, 2839, 2854, 2866, 2873, 2876. On 4 October 1694 the Marquis de l'Hospital
remarked in a letter to Huygens (letter no. 2879): "Je n'ai plus de curiosité de voir ce
qu'il y a de Mr. Neuton dans le livre de Vallis apres ce que vous me mandez."

oversimplified—image that Pierre Brunet has handed on to us. To this end I wish first to discuss briefly the central figure of this reinterpretation: the French philosopher, Father Nicolas Malebranche (1638–1715). Malebranche has aroused a mild amount of interest on the part of English and American historians of philosophy for the influence he exerted upon the thought of John Norris and Bishop Berkeley, on Hume's dismemberment of the notion of causation, and at one remove upon the American Samuel Johnson (1696–1772). For most modern writers, however, Malebranche simply appears as a relic of a religious age, a metaphysician whose goal, unlike that of Descartes to whom he owed so much, was to put the master's New Philosophy at the service of religion, not for ensuring man's dominance over nature. While it is not difficult to describe Malebranche's philosophical position—it has been done many times—it is less easy to categorize him and determine at what points, and how far, he departed from Descartes.[43] His Christian metaphysics can be described as more voluntarist than (in the theological sense) rationalist, and his epistemology is often summed up in the phrase that "we see all things in God." The influence of Saint Augustine was acknowledged by Malebranche himself,[44] but his Platonism is also evident: ideas are not innate in the human mind (in the Cartesian sense); they are imperfect reflections of ideas *in* God. Indeed, it is not too much to say that Malebranche anchored Plato's archetypal ideas in the divine mind. His doctrine of occasionalism, by no means original with Malebranche, is always stressed when his philosophy is summarized: events in the physical

[43]The extent to which Malebranche departed from Descartes in his fundamental doctrines has been much debated. Compare, for example, M. Geroult's article "Métaphysique et physique de la force chez Descartes et chez Malebranche," in the *Revue de métaphysique et de morale*, 54 (1954), 113–134, with the account of Malebranche by Willis Doney in the *Encyclopedia of Philosophy*, V (New York, 1967), 140–144. In physics Malebranche differed from Descartes on the cause of the solidity of bodies, on the laws of impact, the nature of light, and many other points. He emphasized that Descartes's *Principes de la philosophie* must be read with caution, "sans rien recevoir de ce qu'il dit, que lorsque la force et l'evidence de ses raisons ne nous permettront point d'en douter." Cited by Paul Mouy, *Le développement de la physique cartésienne* (Paris, 1934), p. 279.

[44]See Ferdinant Alquié, *Le Cartésianisme de Malebranche* (Paris, 1974), p. 25 and n. 9.

world are not *caused,* but provide the *occasion* for God, constantly conserving his creation,[45] to set in motion laws of nature which we perceive as *rapports,* or relations, between the objects of our experience.

Malebranche came to philosophy, to Descartes, and to mathematics, fairly late in his career. Frail as a child—indeed appearing so all his long life—he was educated at home until 1754–1756 when he studied at the Collège de la Marche under a Peripatetic master. After three years of theological study at the Sorbonne, he began his novitiate in 1660 in the Congregation of the Oratory, a priestly order founded by Pierre Bérulle during the Catholic religious revival in early seventeenth-century France, and which soon was famed as a liberal teaching order. He was ordained in 1664, having devoted himself to Church history, biblical scholarship, and the study of Hebrew. From that time on until his death he lived in the Paris house of the Oratorians on the rue St. Honoré, across the street from the Louvre where the Academy of Sciences, to which he was admitted in his later years, held its biweekly meetings.

As is well known, Malebranche's intellectual inspiration—what has been called his conversion—came from his reading of Descartes's posthumously published *Traité de l'homme* (1664), which aroused his interest not only in physiology and psychology, but in other branches of science and the great scheme of Cartesian philosophy.[46] He cast aside his historical and linguistic inquiries and set out to master all the writings of Descartes. From these, notably the *Géométrie,* stemmed his preoccupation with mathematics, and

---

[45]The originator of the doctrine of occasionalism is often said to be Geulincx of Antwerp (1625–1669), but other followers of Descartes adopted a similar position. Malebranche, in any case, greatly extended Geulincx's doctrine, giving it a central role in his epistemology and his religious philosophy.

[46]Henri Gouhier, *La vocation de Malebranche* (Paris, 1926) is a fine study of this aspect of Malebranche's career. Gouhier (pp. 56–62) pointed out that it was not the *Traité de l'homme* alone that introduced Malebranche to Descartes. The edition of 1664 which Malebranche purchased was that of Clerselier, and included Descartes's *Description du corps humain,* with its unfinished preface stressing the dualistic doctrine of mind and body, as well as writings of Clerselier and other Cartesians which gave Malebranche a conspectus of Descartes's philosophy in all its breadth. Cf. Alquié, *Cartésianisme,* p. 25.

his urge to keep abreast of developments in the sciences. These interests are clearly evident in the first edition (1675) of his most important work: the *Recherche de la vérité*. This work shows his familiarity with discoveries in embryology, microscopy, and the psychology and physiology of perception.[47] He admired Mariotte's work, as exemplifying the role of experiment, and cited Von Guericke's famous experiments. The books in his library tell us still more: on his shelves were the writings of Steno, Bartholinus, Malpighi, Pecquet, Redi, and Swammerdam among other physicians and naturalists. Chemistry, where we note the books of Béguin, Robert Boyle, and Christopher Glaser, played only a small part. But he owned, among physical and astronomical works, Kepler's *Epitome astronomiae Copernicanae*, Huygen's *Horologium*, and La Hire's *Traité de mécanique*.[48]

Of the scientific disciplines, the centrally important one for Malebranche was mathematics. His library was rich in mathematical works: besides the classical authors (Euclid, Apollonius, and the rest) and of course, the *Géométrie* of Descartes, he owned the books of Herigone and Slusius, Oughtred's famous *Clavis*, Franz van Schooten's *Exercitationes* (the chief work that made Descartes's analytical geometry comprehensible) and Isaac Barrow's *Mathematical Lectures*.[49] And Malebranche once described mathematics as "the foremost and fundamental discipline of all the human sciences," excluding, that is, theology, which is a divine, not a human, subject.

At the Oratory, always the teacher, Malebranche brought

[47]For a compact view of Malebranche as mathematician and savant, see the article by Pierre Costabel in the DSB, IX (1974), 47–53. A good introduction to Malebranche's interest in the progress of the life sciences is the single volume of the abortive *Oeuvres complètes de Malebranche*, ed. Désiré Roustan with the collaboration of Paul Schrecker, of which only the one volume (Paris, 1938) appeared before the outbreak of World War II and Schrecker's emigration to the United States. See especially the "Notes des éditeurs," pp. 399–447. Of interest too is Schrecker's "Malebranche et le préformisme biologique," *Revue internationale de philosophie*, 1 (1938), 77–97.
[48]*Oeuvres de Malebranche*, XX ("Malebranche vivant," ed. André Robinet, 1967), chap. 6, "La bibliothèque de Malebranche."
[49]Ibid.

together a group of mathematicians and physicists who are fairly credited with introducing the Leibnizian calculus into France,[50] and this in turn opened the way to the fuller understanding of Newton's accomplishments. With one of his early disciples, Jean Prestet (1648–1690), Malebranche supervised, or at least collaborated in the writing of, an *Elémens des mathématiques* (1675) that was published under Prestet's name. Malebranche was later to disavow this conservative treatise, for it supported the Cartesian position that mathematics has no business dealing with the infinite, whether large or small. Later this group came to include a fellow Oratorian, Father Charles-René Reyneau (1656–1728), Louis Carré (1663–1711), Pierre Rémond de Montmort (1678–1720), and the man whom André Robinet has called the leader, the "chef de file," of the Malebranchistes: the Marquis de l'Hospital (1661–1704). These men not only introduced the Leibnizian calculus into France, but defended it against its conservative detractors, like Pierre Rolle (1652–1749). On the fringes of this group, all of whom sooner or later became members of the Academy of Sciences, stood the figure of Pierre Varignon (1654–1722), something of a late convert to the new mathematics.

The steps by which the calculus came to France have often been retraced, and a brief summary here should suffice. Leibniz's first publication of 1684 on the calculus was a compressed memoir of six pages published in the *Acta eruditorum*. Cryptic, tightly constructed, and further obscured by numerous misprints, it made no immediate impact upon the scientific world. Nevertheless, without instruction or elucidations from Leibniz, Jacob (James) Bernoulli, a professor of mathematics at Basel, fought his way through it, and taught it to his younger brother Johann (John or Jean I). Johann, on a visit to Paris in 1691–1692, came into contact with Malebranche and his circle and lectured on the Leibnizian calculus. His

---

[50]*Oeuvres de Malebranche*, XX, chap. 3, "Le groupe malebranchiste de l'Oratoire," 137–170; André Robinet, "Le groupe malebranchiste introducteur du calcul infinitésimal en France," *Revue d'histoire des sciences*, 13 (1960), 287–308; André Robinet, *Malebranche de l'Académie des sciences* (Paris, 1970).

lessons, as well as notes taken by Malebranche, have survived.[51] Here Johann met the Marquis de l'Hospital, who became his most assiduous convert, studying with him in Paris and engaging him to continue his teaching at l'Hospital's country seat at Oucques, a village in the Orléanais. In 1696 appeared the first French textbook of the calculus, l'Hospital's *Analyse des infiniment petits pour l'intelligence des lignes courbes*.[52]

Before going further, we should perhaps stop to ask what made the Malebranche circle, and Malebranche himself, so receptive to the Leibnizian "infinitesimal calculus," as it soon came to be called. How, to paraphrase André Robinet, can one explain the transformation of Malebranche–Prestet, opposed to infinitudes, into Malebranche–l'Hospital, so readily persuaded to adopt this radical new posture?

Various scholars have pointed out the fact that Malebranche's metaphysical principles dovetailed admirably with his mathematical interests. In particular, it has been suggested that the philosophy of mathematics Malebranche evolved in his later years—which tied in closely with his epistemology, and which can be traced as it evolved in later editions of the *Recherche de la vérité*—was a powerful force leading him to depart from the strictly Cartesian mathematics with which he had begun.[53]

Malebranche's theory of mathematics is intimately tied to his metaphysical postulate that the only truth open to man's finite intellect is the perception of the relations, the *rapports*, between

[51]*Oeuvres de Malebranche*, XVII-2 (*Mathematica*, ed. Pierre Costabel, 1968), 131–294.

[52]Only in the fifth edition of the *Recherche de la vérité* does Malebranche mention the Marquis de l'Hospital and his book, and introduce the names of two new mathematical sciences, the differential and the integral calculus. The former, he writes, has been carefully treated by l'Hospital; the letter still awaits a comparable book, although "plusieurs savants géomètres" are working on the subject. For the moment one must be content with the "petit ouvrage de M. Carré," his *Méthode pour la mesure des surfaces*, etc. Cited by Mouy, *Développement de la physique cartésienne*, p. 269.

[53]See Paul Schrecker, "Malebranche et les mathématiques," *Travaux du IXᵉ congrès international de philosophie—Congrès Descartes* (Paris, 1937), pp. 33–40, and his "Le parallélisme théologico-mathématique chez Malebranche," *Revue philosophique*, 63 (1938), 215–252. Also André Robinet, "La philosophie malebranchiste des mathématiques," *Revue d'histoire des sciences*, 14 (1961), 205–254.

things; knowledge can only be knowledge of such relations. Furthermore, the clearest and most distinct relations we can determine are those of equality and inequality: relations of magnitude. Consequently, since mathematics is precisely the science of these relations, it is the most exact and unimpeachable form of knowledge we can attain. The emphasis here, as in Malebranche's epistemology, must be placed on formal relations, relations of relations, and so on, *instead of on conceivability.* These basic assumptions led to important consequences: (1) doubts were set aside as to the logical foundations of the calculus, widely recognized as dubious, so the way was opened to accept infinitesimals, the inconceivable "infinitely small"; and (2) a premium was placed upon enriching and developing the mathematically expressed laws increasingly used to describe the observed relations of the natural world; (3) finally those, or at least some of those, who accepted these assumptions, were disposed to look with a degree of understanding upon Newton's mathematical world picture.

It would be interesting to know with some exactitude when Malebranche, or members of his group, first came to know Newton's *Principia*. What Paul Mouy calls the "diatribe" against the idea of attraction in the first edition of the *Recherche de la vérité* (1675) could not, from the date alone, have been directed against Newton who, in any case, was invoking the aether in his "Hypothesis explaining the properties of light," and was not (as Mouy has claimed) already an attractionist. We are told that Malebranche first came into direct contact with Newton's writings sometime between 1700 (when the fifth edition of the *Recherche* appeared) and the year 1712, when the sixth edition of that work was published.[54] This may be so, for Malebranche read Newton's *Opticks* soon after its appearance and became at least a partial convert to Newton's discoveries about color. Yet he certainly knew about the *Principia*, and heard it discussed, at a much earlier date. A letter from Malebranche's friend Jacquemet to Charles-René Reyneau, written on 9 April

[54]See Paul Mouy, "Malebranche et Newton," *Revue de métaphysique et de morale,* 45 (1938), 411–435.

1690, contains the remark—if indeed it was not a boast—that he, Jacquemet, had completed his reading of Newton's book earlier that year.[55] Yet, although there is no mention of the *Principia* in Malebranche's *Recherche*, there is a passage in the sixth edition of that work (the edition where his approval of Newton's optical experiments is announced) which suggests that he had grasped Newton's essential discovery concerning planetary motion. For what the evidence is worth, we know that a copy of the first edition of the *Principia* is listed as having had a place on Malebranche's bookshelves.

It was soon evident to the men of Malebranche's circle that Newton's celestial dynamics—his mathematical approach to physical nature—had to be taken seriously, despite his attack on Descartes's vortex theory of planetary motion, from which they could not free themselves completely. There was much in the *Principia* with which they were intellectually attuned: newly discovered relations or *rapports*, laws of the physical world, the mathematical approach to nature. It struck a number of them that this Newtonian mathematization of nature could be clarified and enriched if it could be translated into Leibnizian analytical methods and symbolism.

Both Johann Bernoulli and the Marquis de l'Hospital began cautiously to treat central forces analytically. But it was Varignon, Professor of Mathematics at the newly founded Collège Mazarin (a post he later combined with a chair at the Collège Royal), who took up these questions as matters of real importance. Earlier he had remained faithful to Descartes and had difficulty in moving beyond a mathematics of finitude to an acceptance of the Leibnizian calculus. But at last, reassured by Malebranche's position on mathematical intelligibility (as distinguished from conceivability), he became a convert to the new mathematics and turned to applying the calculus of Leibniz to Newtonian dynamics.[56] In a series

[55]*Oeuvres de Malebranche*, XVII-2 (*Mathematica*, ed. Pierre Costabel, 1968), 62. See also the later letter in which Jacquemet thanks Reyneau for information on Barrow's method, remarking that "dans le fond" it is the same as that of the Marquis de l'Hospital and Newton, except that the latter applied it to incommensurables "qu'on prétend être une des plus belles et des plus utiles inventions de ce siècle dont Messieurs Leibniz et Newton ont tout l'honneur" (ibid., p. 61).

of papers contributed to the *Mémoires* of the Academy of Sciences during the early years of the eighteenth century he treated such problems as: (1) given the law of force, to find the path of a moving body; or (2) conversely, given the path, to find the implied law of force. He was able to show, for example, that a logarithmic or hyperbolic spiral path implied a central force proportional to the cube of the distance.

Earlier in this paper I questioned the sharp polarity between Cartesians and Newtonians that Voltaire taught us and which, in our century, Pierre Brunet set forth in his book on the acceptance of the Newtonian physics in France. I shall now suggest that the followers of Malebranche—indeed the aging philosopher him-self—occupied a midway position, and conceded enough to New-ton to pave the way for the full-fledged Newtonianism of the later eighteenth century. These men, far from merely tinkering with Descartes's *tourbillon* model (which they continued to do as had Huygens and Leibniz), abandoned at least the outworks of the Cartesian fortress. These Malebranchistes in the Royal Academy of Sciences brought into the Age of Enlightenment not only Male-branche's doctrine that knowledge of nature is a knowledge of mathematical relationships, but carried too the Master's respect for evidence, his willingness to recast, or depart from, the specific doctrines of Descartes. In so doing they weakened the opposition to Newton and prepared the way for the more militant Newtonian-ism of Maupertuis, Clairaut, and Voltaire in the late 1730s. Let me offer the following general propositions to be examined *à fond* by others:

(1) These Malebranchistes (and in this they were not alone) recog-nized Newton as a major figure to reckon with, not only in mathematics, but in the mathematizing of nature.

[56]For Varignon, see the article by Pierre Costabel in DSB, XIII (1976), 584–587, and his *Pierre Varignon et la diffusion en France du calcul différentiel et intégral* (Paris, 1965). An important article is J. O. Fleckenstein, "Pierre Varignon und die mathe-matischen Wissenschaften im Zeitalter der Cartesianismus," *Archives internationales d'histoire des sciences*, 2 (1945), 76–138.

(2) On the basis of what they regarded as Newton's experimental brilliance, they were the first in France to deem Newton's optical experiments a model of proper procedure to follow, and an essential part of his doctrine of the origin of color, albeit with certain theoretical deviations.

(3) They were won over to Newton's laws, the inverse square principle of "universal" gravitation, and the successful application of his mathematical rules of nature to Kepler's empirical description of planetary motion.

All this they accepted without being obliged to adopt the notion of a void, and the apparently absurd notion of bodies attracting one another through the emptiness of space. Although all adhered to some sort of doctrine of an aetherial vortex, the concessions, the departures from Descartes, paved the way for an overt acceptance of Newtonian doctrine, root and branch.

The first steps were taken by Malebranche himself when in several respects he departed from Cartesian physical thought. Most thoroughly studied are the changes he made in the Cartesian laws of impact—probably influenced by the experiments of John Wallis and of Mariotte—but he further denied that cohesion and solidity were simply caused by matter at rest. Indeed, in suggesting that rest is a mere privation of motion, he came close to the Newtonian position that rest and motion are merely "states" of matter. More fundamental, perhaps, was his modification of Descartes's theory of matter.[57] Descartes constructed his universe of three elements: his first element is composed of exceedingly fine particles that make up the luminous matter of the sun and fixed stars, and are so fine, and of such varied shapes, that they can fill all the spaces between the other elements. The second element is the key to Descartes's physical model; it consists of rounded, hard, and inflexible particles through which the light of the sun is instantaneously transmitted to illuminate the larger masses of the terrestrial bodies: the earth, moon, the planets and their satellites, that make up the third element.

Malebranche, even before encountering Newton, had fun-

[57]Mouy, *Développement de la physique cartésienne*, pp. 282–290.

damentally revised this picture of the second element. Instead of being hard spheres, the particles are compressed fluid matter, tiny vortices. Nor did he precisely follow the Cartesian theory of the nature of light. For example, Olaus Roemer's discovery convinced him that light was not transmitted instantaneously, but required time for its passage through the second element.

Soon after his election to the Academy of Sciences in 1699, Malebranche communicated a theory of light and color that departed from Descartes. Here he attempted to supply a mechanical model, a plausible hypothesis, to account for the origin of color. Light, to Malebranche as to Pardies, Huygens, and others, is a pulse or vibratory motion of the aether or second element. But for Malebranche, colors result from the different frequencies of these pulses. Not all the hues of the spectrum are produced by this mechanism, but only the *primaries* (red, yellow, and blue). The sensation of white is produced when the original vibrations from a luminous source are unaltered in their transmission.[58] This is a version of what historians call a *modification* theory of prismatic color.

At the time his paper was presented, Malebranche seems to have been wholly ignorant of, or indifferent to, Newton's famous early paper of 1672.[59] In 1706, it seems, the Latin edition of Newton's *Opticks* came into his hands, and he was soon converted. Thus when he prepared the last edition of his *Recherche de la vérité*, the sixth (1712), he profoundly altered that appendix or "éclaircissement" he had added in 1700 and which was, in effect, his 1699 paper. In his revision he accepted Newton's notion that there existed in white light an infinite range of properties producing the different colors. The colors corresponded to various frequencies which the prism sorted out. Each color, he wrote, citing the "excellent work of M. Newton," has its characteristic refrangibility. He even adopts Newton's terminology, at least up to a point. The "primitive colours" of his earlier paper he now refers to, as Newton does, as "simple" or

[58]Pierre Duhem, "L'optique de Malebranche," *Revue de métaphysique et de morale,* 43 (1916), 37–91.
[59]Ibid.

"homogeneous." White solar light ("the most composite of all") is described according to his own adaptation of Newton as "composed of an assemblage of different vibrations." Owing in part to Malebranche's great influence, Newton's *Opticks* was accepted, early in the eighteenth century in France, as a model of scientific inquiry.[60] One of Malebranche's disciples, Pierre Varignon, was a key figure in seeing through the press the Paris edition (1722) of Pierre Coste's French translation of the *Opticks*.[61] But it should be emphasized that there is no evidence that Malebranche, an elderly cleric, repeated any of Newton's optical experiments. He seems simply to have been persuaded by Newton's testimony as presented in the *Optice*. This last edition of the *Recherche* shows that Malebranche was led on to the *Principia,* for he added a sort of appendix to his theory of little vortices to alter the Cartesian explanation of gravity and adjust his thought so far as possible to Newton's findings.

To illustrate my point that we can no longer justly speak of an academic world divided between Cartesians and Newtonians (or, if you will, dominated by strict Cartesians struggling to fend off a Newtonian invasion) I should like to say something about three of Malebranche's disciples and admirers: the Oratorian father, Charles René Reyneau, Jean-Baptiste Dortous de Mairan (1678–1771), and Joseph Privat de Molières (1676–1742).

In 1708 Father Reyneau, a professor of mathematics at Angers, published his *Analyse démontrée* which continued and brought up to date Descartes's work on the theory of equations and introduced the reader to the calculus. Modern analysis, he wrote, had its beginnings in Descartes's *Géométrie*. But Descartes lacked a mathematics that could precisely represent nature; for nature produces curves by continuous motion, and nature's curves are made up of "parties insensibles," parts smaller than we can determine, and of instants of time swifter than we can imagine. What was needed was a new form of mathematical expression, and this was devised "at the same

---

[60]As, for example, Fontenelle's *éloge* of Newton. For the early English version (London, 1728) see *Newton's Papers*, pp. 444–474.
[61]Hall, "Newton in France," p. 244.

time in Germany by M. Leibniz and in England by M. Newton."
Further on he mentions the "savant ouvrage" (the learned work) of
Newton, the *Principia*.[62]

Dortous de Mairan, a young provincial of the country gentry,
was born in Béziers. He was sent to Paris to attend a sort of
finishing school, where horsemanship, fencing, dancing, and other
proper accomplishments were taught to young noblemen. In Paris,
he came under the spell of Malebranche who taught him to understand the Marquis de l'Hospital's *Analyse des infiniment petits,* and
gave him other lessons in mathematics and physics.[63]

Mairan returned to his native Languedoc. Uncertain of his future course, he toyed with the philosophical and theological difficulties encountered in reading Spinoza, and wrote his old teacher,
the aging Malebranche, for guidance.

In their correspondence, Malebranche not only rescued Dortous
from the pitfall of Spinozistic pantheism, but incidentally informed
his young friend that he had abandoned his earlier views about
light and color after reading Newton's *Opticks*. He went on to tell
Mairan (in August 1714) that his version of Newton's theory could
be found in the last edition of his *Recherche*. Mairan promptly acquired Malebranche's book, and probably soon afterward the Latin
Optice as well, for Pierre Coste, the translator of the *Opticks* into
French (1720 and 1722), tells us that Mairan in 1716–1717 was the
first in France to repeat successfully Newton's optical experiments.
The influence of Newton's work soon became apparent in Mairan's
memoirs. In his earliest scientific contribution, his *Dissertation sur les
variations du baromètre* (Bordeaux, 1715), he mentioned Richer's
discovery that a pendulum, whose period of oscillation in Paris is a
second, has to be shortened near the equator. And he remarked
that "certain celebrated mathematicians" had concluded that the
earth was a globe flattened at the poles. A note cites Huygens's

---

[62]Mouy, "Malebranche et Newton," p. 421.

[63]See my "Newtonianism of Dortous de Mairan," reprinted in my *Essays and Papers in the History of Modern Science* (Baltimore and London, 1977), pp. 479–490. Origi-nally published in the Festschrift for Ira Wade, it suffered from some typographical legerdemain on the part of the printer. This has been corrected, and the article somewhat expanded.

Discours de la cause de la pesanteur and Newton's *Principia*.[64] The following year (1716) in his *Dissertation sur la glace* Mairan showed himself quite at home with the *Optice,* for he wrote about the relation of color to refrangibility, the transparency of bodies, and other matters discussed by Newton.

The most explicit of Mairan's early references to Newton's work on light and color appears in a short essay on phosphorescence published in 1717.[65] Here he summarized Newton's discovery that each colored ray has its characteristic refrangibility, and recounted the "ingenious experiments" which had led Newton to this discovery "in order to acquaint those who have not seen the *Opticks* of Mr. Newton with what I shall have to say on this matter."

Not so long after, we find Mairan in Paris, for in 1718 he became *associé géomètre* in the Academy of Sciences, taking the place of Guisnée, a member of Malebranche's circle who died in that year. His earliest contributions as an academician show the profound influence the *Opticks* had on him. In one, he draws upon Newton's analogy between the colors of the spectrum and the intervals of the octave and suggests that, just as Malebranche believed colors to be caused by differently vibrating globules of the *matière subtile,* so air must transmit sound by means of distinct particles having different rates of vibration.[66] In another paper he argued against Descartes's theory that colors are derived from the differential rotation of kinds of globules, by showing that spheres rotating differently and striking a surface obliquely would reflect at different angles. Thus, contrary to the established law of equiangular reflection, rays of different colors would then have characteristic reflectivities.[67] Mairan, more explicitly than Newton, held an emission theory in

[64]Mairan had doubtless read Huygens's "Discours de la cause de la pesanteur" appended to his *Traité de la lumière.* Whether at this time he had seen Newton's *Principia* is less certain, although he refers to it, for he could have learned of Newton's views on the shape of the earth from the remarks in Huygens's "Addition." See above, n. 31.

[65]*Dissertation sur la cause de la lumière des phosphores et des noctiluques* (Bordeaux, 1717), p. 48.

[66]Fontenelle, *His. Acad. Sci.,* 1720 (1722), pp. 11–12 (cited by Brunet, pp. 84–85). See also the article on Mairan (by Sigalia Dostrovsky) in DSB, XIII, 33.

[67]*Mém. Acad. Sci.,* 1722 (1724), pp. 6–51.

which light rays consist of trains of corpuscles, of *corps lumineux*. Moreover, he believed that reflection does not occur at the point of actual contact of rays of light with a solid surface, but when the corpuscles encounter a "fluide subtile répandu dans leurs pores," a view he recognized as similar to that held by Newton in the *Opticks*.[68]

There can be little doubt that Mairan was reluctant to abandon the mechanism of subtle fluids, nor that he felt increasingly allured by Newton's views and by his own respect for experimental evidence. Often he was troubled by contradictions in empirical data and he tried to reconcile them. For example he found Jacques Cassini's geodetic measurements, leading to the inference of an elongated earth, in conflict with Huygens's and Newton's inference from Richer's pendulum observations at Cayenne that the earth should have the shape of an oblate spheroid. He attempted to show that both observations led to a *sphéroïde oblong*, if one denied the primitive sphericity of the earth which Huygens had assumed as a postulate.[69]

Mairan's reluctance to give up the vortex theory led him to some extraordinary mental gymnastics. A much-cited argument against Descartes's *tourbillons* was the existence of retrograde comets, comets whose motion is opposite to that of the planets. Mairan pointed out that planets at portions of their orbits appear to stop and reverse their "direct" motion from west to east, but he pointed out that this is merely an optical effect produced by a combination of the earth's revolution and the direct planetary path. Could not retrograde comets be planets that are only visible during the retrograde portion of their path?

Unlike some of his contemporaries who favored the existence of a deferent subtle fluid, Mairan had—even more than Malebranche—a keen respect for observed fact. Certain Cartesians, like Villemont, argued that comets did not enter the solar system below Saturn. Mairan knew that in fact, on occasion, they did.

[68]Ibid., pp. 50–51. Cited by Brunet, pp. 115–116.
[69]"Recherches géométriques sur la diminution des degrés terrestres, en allant de l'équateur vers les pôles," *Mém. Acad. Sci.*, 1720 (1722), pp. 231–277.

But if they come close, yet do not penetrate our solar vortex, then our vortex, Mairan argued, cannot be spherical but must sometimes be depressed. Comets, he suggested, are planets of nearby vortices, moving about their own suns; their vortices can act upon others, including ours, engaging one another, as Fontenelle put it—like gears of a clock—but altering their shape. The flattening of our vortex allows comets to approach closely yet without penetrating it.[70]

We can detect, I believe, a growing familiarity with, and acceptance of, much that we associate with Newton. In a memoir of 1724 he sought to explain short-range forces of attraction and repulsion, in particular the behavior of water and mercury in capillary tubes. Assuming that around all bodies—not only magnets—there is an atmosphere of *matière subtile,* he sought to explain why bodies attract or repel. Water wets glass because the atmospheres are in some manner compatible, whereas in the case of mercury its surrounding atmosphere is opposed by that of the glass, thus accounting for the convex meniscus. Here, as Fontenelle perceived, Mairan was toying dangerously with those attractive forces which a good Cartesian held in abhorrence.[71]

Another problem brought the Cartesian scheme into court, that of the diurnal rotation of the earth. The Newtonian world view simply took the spin of the earth for granted: once established, inertia would keep the planets rotating. But there was no way that the Newtonian attractionist system could start a planet spinning. A Cartesian vortex by itself was little help: indeed it would seem to demand a rotation opposite to that which actually takes place. Mairan believed the key was the assumption that the two hemispheres of a spherical planet must "weigh" differently towards the sun, according to the inverse square law of gravity, and respond oppositely to the deferent fluid of the vortex. Brunet found it

[70]Brunet, pp. 134–135.
[71]Brunet writes: "Cette explication par la physique tourbillonnaire est d'autant plus caractéristique ici des préférences cartésiennes de Dortous de Mairan que, puisqu'il faisait appel à une sorte d'extension du magnétisme, il pouvait encore trouver là une occasion de se rallier, plus ou moins directement et explicitement, à la théorie de l'attraction" (ibid., p. 121).

ironic that a Cartesian should find it necessary to go to Newton for principles with which to defend Cartesianism.[72]

There is clear evidence that Mairan was moved to study the *Principia*. From Newton's gravitational data, he estimated the relative weight of an identical mass on the surface of the sun and on the earth, noting that the numbers Newton cited differed in the three editions, because—he pointed out—Newton used different values for the solar parallax. By 1733 Mairan had not only made use of points here and there in the *Principia* but had gone a long way toward accepting Newton's celestial dynamics, and the Newtonian principle that central forces in the solar system operate according to the inverse square law. These laws of the solar system, he wrote, are well known and fit modern observations. And he continues:

> We therefore admit these principles in conformity with what one finds about them in the *Mathematical principles* of Newton . . . without claiming to enter . . . into the discussion of causes.
>
> The heavens better understood, the laws of motion better developed, gave to this great man [Newton] an advantage over Descartes and the early Cartesians which cannot deprive them of the glory they have justly gained . . . or forbid them the use of knowledge that time has brought forth, on the pretext that this knowledge did not emanate from their school.[73]

My third specimen, Joseph Privat de Molières (1677–1742), was, like Dortous de Mairan, a man from the Midi. Born in Provence of a distinguished family he was educated in various nearby Oratorian schools (at Aix-en-Provence, Arles, and Marseilles), eventually ending up at Angers, where he studied mathematics and natural philosophy with Charles René Reyneau in 1698–1699. This relationship influenced his future. Against the wishes of his family, he

---

[72]Ibid., p. 170. For a detailed analysis of Mairan's theory of planetary notation, see Aiton, *Vortex Theory*, pp. 182–187.

[73]*Traité physique et historique de l'aurore boréale* (Paris, 1733), p. 88. In the expanded edition of his *Dissertation sur la glace* (Paris, 1749), Dortous de Mairan expressed his pleasure that Newton's letter to Boyle of February 1678/79, recently published by Thomas Birch in his *Life of the Honourable Robert Boyle* (1744), showed Newton an advocate of the sort of *matière subtile* that he, Mairan, used to explain various phenomena. See the "Preface," pp. xviii–xxii.

became a priest of the Congregation of the Oratory, and in 1704, determined to devote his life to science, he came to Paris to sit at the feet of Malebranche and absorb his wisdom.[74] To Malebranche's influence we can safely attribute his central concern: the elaboration of a modified Cartesian physics in which, while remaining faithful to the notion of a *plenum* and to strictly mechanical explanations, he attempted to account mathematically for the inverse square law of gravitation and Newton's demonstration of Kepler's laws, by means of a modified vortex model.

In 1721 Privat de Molières entered the Academy of Sciences as *adjoint mécanicien,* and two years later he succeeded Varignon as professor of philosophy at the Collège Royal. In memoirs read to the Academy in 1728 and 1729, and at greater length in the four volumes of his *Leçons de physique*—lectures delivered at the Collège Royal and published between 1734 and 1739—he concentrated upon what Mme du Châtelet called the "curious business" of trying to reconcile Newton and Descartes. The result was what Robinet has called a "monument malebranchiste," in which—while opposing the doctrines of empty space and the hypothesis of attraction— he freed himself from the narrow Cartesians and took the position of the "cartésiens malebranchistes réformateurs."[75]

In 1728, taking his cue from Varignon's work on central forces, he launched his effort to shore up the theory of *tourbillons* by a mathematical analysis of the centrifugal forces at work in a cylindrical vortex whose axis is equal to the diameter of its base. He found that if the various layers into which one imagines the cylindrical mass to be composed revolve in times proportional to the distance from the axis, no part or globule will approach or recede from the axis. On the other hand, in a *spherical* vortex the condition

---

[74]The primary source for biographical information on Privat de Molières is the *éloge* pronounced by his friend Dortous de Mairan in *Hist. Acad. Sci.,* 1742 (1745), pp. 195–205, reprinted in Jean-Jacques Dortous de Mairan, *Eloges des académiciens de l'Académie royale des Sciences, morts dans les années 1741, 1742, 1743* (Paris, 1747), pp. 201–234. There is a brief summary by Martin Fichman in his article on Privat de Molières in DSB, XI (1975), 157–158.
[75]*Oeuvres de Malebranche,* XX, 170–171.

of equilibrium is different: there will be stability of all layers if the central forces are inversely as the square of the distance from the center.[76] Privat de Molières claimed to show, also, that in a spherical vortex the distances from the center of points moving in the concentric shells are as the cube roots of the squares of the periodic times. This, he pointed out, was "la fameuse règle de Képler."[77] Unfortunately, as he recognized, this held only for points moving in the plane of the equator, and his law lacked the generality it had with Kepler and Newton.

In 1729 he tackled the problem of deriving Kepler's First Law, i.e., the elliptical path of planets, from vortex theory.[78] It was clear that the problem was insoluble if the vortex was made up of the hard globules of Descartes. In consequence, he adapted Malebranche's theory of *petits tourbillons*, of small, elastic vortices, composed in their turn of still smaller vortices (*tourbillons du second genre*).[79] In effect he assumed the infinite divisibility of matter, imagining as many levels as were necessary to account for particular phenomena, and applying to matter the *infiniment petits* so useful to the "mathematicians of our age." His model envisaged a planet moving in a vortex "distorted into an elliptical shape by the unequal pressures of the neighbouring vortices."[80]

In the *Leçons de physique* his theory of mini-vortices was elaborated and applied not only to celestial mechanics, but also to chemistry and electricity. The successive volumes of the *Leçons* received the official approval of committees of the Academy of Sciences, on each of which sat his Malebranchiste colleague, Dor-

[76]"Lois générales du mouvement dans le tourbillon sphérique," in *Mem. Acad. Sci.*, 1728 (1730), 245–267.

[77]Cited by Brunet, p. 159.

[78]"Problème physico-mathématique, dont la solution tend à servir de réponse à une des objections de M. Newton contre la possibilité des tourbillons célestes," in *Mém. Acad. Sci.*, 1729 (1731), pp. 235–244.

[79]Privat de Molières was not alone in being influenced by Malebranche's theory of *petits tourbillons*. In 1726 Pierre Mazière offered a mechanical explanation of elastic collision in terms of such tiny aetherial vortices. See Carolyn Iltis, "The Decline of Cartesianism in Mechanics: The Leibnizian-Cartesian Debates," *Isis*, 64 (1973), 360–363. Cf. Brunet, *L'introduction*, pp. 140–144.

[80]Aiton, *Vortex Theory*, p. 209.

tous de Mairan.[81] From the first, Privat de Molières set out to demonstrate, as he put it, that "the chief doctrines of the two most
celebrated philosophers of our time, Descartes and Newton,"
which appear so incompatible, can in fact be reconciled. In a series
of closely articulated propositions, the reader, he predicted, will
find "perhaps with surprise that, although the two men followed
what seem to be completely opposing paths, these paths nevertheless led to the same goal." And Privat goes on confidently:

> You will see emerging out of the plenist system that Descartes fol-
> lowed, even Newton's void, that non-resisting space of which this phi-
> losopher has irrefutably established the existence.

And from impulse or impact he will find derived that attraction of
gravity

> which increases and decreases inversely as the squares of the dis-
> tances, from which Newton, without however being able to discover
> the mechanical cause, has drawn so many splendid consequences,
> based on a calculus . . . of which this great man is the first inventor.[82]

Dortous de Mairan and Privat de Molières have both been described as leading eighteenth-century Cartesians. Pierre Brunet
gives special treatment to both men in his chapter entitled "L'effort
des grands cartésiens." Quite recently Martin Fichman wrote in his
article in the *Dictionary of Scientific Biography* that Dortous de
Mairan was a "major figure in the protracted struggle against the
importation of Newtonian science in France," and "devoted his
career to developing and improving Cartesian physics."

[81]For vol. I (1734) the committee was composed of Mairan and Louis Godin; for
the subsequent three volumes of 1735, 1737, and 1739, the committee consisted of
Mairan and the Abbé de Bragelongne, the latter also a disciple of Malebranche. See
Robinet in *Oeuvres de Malebranche*, XX, 152–153, 170, 359.

[82]Joseph Privat de Molières, *Leçons de physique*, 4 vols. (Paris, 1734–39), I, vii–x. His
vortices, he writes (I, 307), provide "une cause mécanique de la pesanteur, ou de la
force centripète, telle que M. Newton la demande, qui croît & décroît en raison
inverse des quarrés des distances au centre, & qu'il avouë n'avoir pû déduire de ses
suppositions." Newton, he adds (I, 308), was obliged to regard gravity as a universal
principle "& un effet sans cause."

Yet both writers—Brunet and Fichman—have had to emphasize that the two men in question were by no means rigidly Cartesian. At one point, Brunet remarked that while Dortous remained attached to Cartesianism on most fundamental questions (whatever that may mean) he was nevertheless enticed—his word is *séduit*—by Newton's ideas, chiefly because of his admiration, like Fontenelle's, for Newton's experimental skill. And Martin Fichman conceded in another *Dictionary* article that Privat de Molières, while persuaded of the correctness of Descartes's ideal of a purely mechanical science (that is, of a mechanics of impact) was nevertheless "cognizant of the superiority of Newtonian precision in comparison with Cartesian vagueness in its explication of natural phenomena." Vagueness is hardly a word I should use in connection with Descartes. I was glad to discover that my friend John Heilbron of Berkeley, in his book on the early history of electricity, quite bluntly and accurately calls Privat de Molières "a devout Malebranchiste."

I can only repeat the main theme of this part of my paper: that Malebranche and his followers broke down the initial barriers of the Cartesian fortress, and made the way easier for radical Newtonians like Maupertuis, Clairaut, and Voltaire.[83] Paul Mouy did not exaggerate when he wrote that around 1730 an eclectic fusion of Malebranchiste and Newtonian ideas was very much *à la mode* in French science. As Voltaire's Minerva of France, his "immortelle Emilie," wrote of Cartesianism to Cisternay Dufay: "It is a house collapsing into ruins, propped up on every side . . . I think it would be prudent to leave."[84]

[83]For the persistent influence of Malebranche, especially his criticism of the concept of force, on these later Newtonians, notably Maupertuis, see the excellent article of Thomas L. Hankins, "The Influence of Malebranche on the Science of Mechanics during the Eighteenth Century," *Journal of the History of Ideas*, 28 (1967), 193–210.

[84]*Les lettres de la marquise du Châtelet,* ed. T. Besterman, 2 vols. (Geneva, 1958), I, 261. Cited by J. L. Heilbron in his *Electricity in the 17th and 18th Centuries: A Study of Early Modern Physics* (Berkeley and Los Angeles, 1979), p. 278.

CHAPTER 4

Newton in France

Two Minor Episodes

Pierre Brunet has made it clear that not until the 1730s did Newtonian physics receive influential acceptance in France.[1] Whether accurate or not, the impression Brunet conveys is that before this time it was Newton's work on light and color, not the massive achievement of the *Principia,* which evoked the admiration of his French confrères. Yet despite his detailed discussion of the "résistance cartésienne" before 1738, Brunet's picture is far from complete. With this note I merely wish to call attention to two minor episodes about which Brunet is silent.

I

In 1718 there took place an exchange of letters between two anonymous French naval officers as to the merits of Newton's *Principia.* One of these *capitaines de vaisseau* had become converted to the new ideas, after a careful reading of the great work, for he wrote his friend that in his opinion "Monsieur Newton avoit *coulé à fond* Monsieur Descartes." His correspondent was of a different opinion, so to settle the argument the two men agreed to submit to

From *Isis,* 53 (1962), copyright © 1962 by the History of Science Society, Inc., and reprinted here with permission.
[1]Pierre Brunet, *L'introduction des théories de Newton en France au XVIIIᵉ siècle,* I: Avant 1738 (Paris, 1931). See also his *Maupertuis—Etude Biographique* (Paris, 1929).

the judgment of a Jesuit astronomer, a man who, in all likelihood, had been the teacher of one or both of these officers.

It is from this referee, Father Antoine de Laval (1664–1728), that we learn of this little-known debate.[2] The discussion of the two officers, Father Laval tells us, made a greater impression on him than his own reading of the *Principia*; accordingly, he set down his reflections for their benefit. These papers Laval later printed as appendices to his *Voyage de la Louisiane* (1728). The first is called "Réflexions sur quelques points du Sistème de M. le Chevalier Newton." The second, written after 1722, refers to Newton's *Opticks.*[3]

Father Laval's prefatory remarks in the "Réflexions" are of considerable interest. It was his purpose, he wrote, only to show that Newton's principles "ne sont pas tous des veritez Mathématiques, qui ne souffrent aucune réplique: & que les Cartesiens peuvent expliquer aussi-bien les Phénomenes de la nature selon leurs hypothèses." While clearly leaning toward Descartes, he is fair and cautious; the arguments of the pro-Newtonian *capitaine de vaisseau* evidently had given him pause. Yet he has his doubts: "Quelque lumineuse que soit la nouvelle Physique, elle a . . . ses obscuritez. . . . Cela me doit rendre plus retenu à prendre parti."

These little-known documents deserve careful study. I mention them here, not only to remind Newtonian scholars of their existence, but also to call attention to a minor, yet perhaps significant, point. Laval confesses that, though he knew the *Principia*, in 1718 he had not read the *Opticks*. "Je ne sçavois point pour lors que l'optique de Monsieur Newton eut été traduite de l'Anglois; ainsi

[2]Antoine de Laval was born in Lyons, 26 October 1664. His life was spent at the two great naval centers of Marseilles and Toulon. At Marseilles, according to Lalande, he had an observatory which he directed for many years. He contributed many observations to the *Mémoires de Trévoux*. He was a close friend and associate of Jean-Mathieu de Chazelles, Professor of Hydrography at Marseilles. Laval may also have taught at Marseilles, but on the title-page of his only book he is described, as "Professeur Royal de Mathematiques, & Maître d'Hydrographie des Officiers & Gardes de la Marine du Port de Toulon." He died 5 September 1728.

[3]*Voyage de la Louisiane, fait par ordre du Roy en l'année mil sept cent vingt: Dans lequel sont traitées diverses matières de Physique, Astronomie, Géographie & Marine . . . Et des Réflexions sur quelques points du Sistème de M. Newton. Par le P. Laval de la Compagnie de Jésus* (Paris, 1728). See pp. 153–191.

pour mon malheur je ne l'avois pas lûë." If this remark refers, as it seems to do, to the time when he was in correspondence with his officer friends, the translation he mentions must obviously be in the Latin *Optice* of 1706, for the French translation by Pierre Coste did not appear until 1720. It would be interesting to know more than we evidently do about the reception of the Latin *Optice* in France. Brunet asserts that this work was "d'usage courant." Is this true? How extensive was its influence?

II

Professor I. Bernard Cohen has discovered in Paris an incomplete or abridged manuscript translation into French of the *Opticks* of 1704.[4] Long antedating the translation by Coste, it has been attributed to the physician and chemist, Etienne François Geoffroy (1672–1731), whom we know to have spent some time in England, to have been a friend of Hans Sloane, and to have been made F.R.S. in 1698. It may be of interest therefore to report that the Cornell University Library has acquired Geoffroy's presentation copy of the first edition of Newton's *Opticks* signed by the recipient in the year 1704. Cornell's copy of this famous quarto is in a good state of preservation, though the margins have been slightly cropped. On the upper right-hand margin of page 1 we read the faded inscription: "Pour Mr Geoffro[y] de l'acadé[mie] des Scien[ces]." On the title page in black ink appears the signature "Geoffroy 1704." Elsewhere in the book an owner, almost certainly Geoffroy, has made a lone annotation: on the left-hand margin of page 134 appears the single word "zinck," evidently to explain the word "Spelter" which occurs in the corresponding line of Query 10.

The book was not sent directly by Newton or inscribed by him personally; the inscription, in a formal copper-plate hand, was certainly not written by Newton. In all probability the book was dis-

---

[4]Professor Cohen has published an article in which he discusses this early attempt to render the *Opticks* into French (see Chapter 5, n. 64).

patched by Hans Sloane, the Secretary of the Royal Society, with whom Geoffroy was in frequent correspondence.[5]

The extent of Geoffroy's familiarity with Newton's work is a matter of some importance. Geoffroy read in 1718 the famous memoir which is properly taken as the starting point of an interest in chemical affinity. Was he aware that in a long chemical query of the Latin *Optice* of 1706 Newton had described a series of displacement reactions, and explained them by the theory of differential attractions? In his paper of 1718, Geoffroy makes no reference to Newton, and avoids the use of the word "attraction." Instead, he uses the vague term "rapports" favored, until the end of the century, by most French chemists. There was nothing in the *Opticks* of 1704 to influence Geoffroy in his study of affinity, for it did not contain this important chemical query.[6] Was Geoffroy, like Father Laval, unaware of the Latin *Optice* of 1706? In view of his close connection with the Royal Society, his contact—albeit indirect—with Newton, and the interest he seems to have displayed in the *Opticks* when it first appeared, this is unlikely. It may well have been the appearance of the *Optice* in 1706 which caused him to cast aside the French translation he had undertaken, if indeed it is his work. Any specific evidence of Geoffroy's familiarity with the Latin *Optice,* and with the famous 31st Query, would be of very great interest.

---

[5]In the Sloane MSS in the British Museum are numerous letters of Geoffroy to Hans Sloane. Some use has been made of them by Jean Jacquot: *Le naturaliste Sir Hans Sloane (1660–1753) et les échanges scientifiques entre la France et l'Angleterre* (Conférences du Palais de la Découverte, Série D, No. 25).

[6]Printed for the first time as Query 23 of the Latin *Optice,* it is better known in its expanded form as Query 31 of the second English edition. The passages relating to displacement reactions appear in both editions.

CHAPTER 5

Newton in France

The Delayed Acceptance of His Theory of Color

"**O**ne advantage of this book," Fontenelle wrote of Sir Isaac Newton's *Opticks,* "equal perhaps to the many new discoveries with which it abounds, is that it furnishes us with an excellent model of proceeding in Experimental Philosophy. When we are for prying into Nature, we ought to examine her like Sir Isaac, that is, in as accurate and importunate a manner."[1] Clearly, when Fontenelle's famous *éloge* of Newton was read to the Royal Academy of Sciences in Paris on 12 November 1727, several months after Newton's death, the English scientist's discoveries concerning light and color were valued at their true worth by the French scientific community of which Fontenelle, in this eulogy, as in the others he wrote, was the official spokesman.

A draft of this paper was written in the spring of 1973, but the project was set aside until I could benefit from the letters, notably about the publication of the French translations of the *Opticks,* destined to appear in the last volume of *The Correspondence of Isaac Newton.* Some references to this early draft were made by Professor A. Rupert Hall in his "Newton in France: A New View," *History of Science,* 13 (1975), pp. 233–250. My draft had been sent to him early that same year in the hope that it might prove of use in his task of editing Volume VII of the Newton *Correspondence.*

[1] *The Elogium of Sir Isaac Newton by Monsieur Fontenelle, Secretary of the Royal Academy of Sciences at Paris* (London, 1728), reprinted in facsimile in I. Bernard Cohen, ed., *Isaac Newton's Papers and Letters on Natural Philosophy and Related Documents* (Cambridge, Mass., 1958), pp. 444–474. Compare J. E. Montucla's verdict, probably echoing Fontenelle: "Enfin parut son Optique ce livre admirable, & si digne d'être conseillé à tous ceux qui cultivent la Physique, comme le plus parfait modèle de l'art de faire les expériences" (*Histoire des mathématiques,* 2 vols. [Paris, 1758], II, 623).

The Delayed Acceptance of His Theory of Color

Newton's optical investigations had not always enjoyed such esteem, but our knowledge of how his experiments were first received on the Continent, how they were apparently refuted, how they were finally and belatedly confirmed, is fragmentary to say the least, and perhaps justifies this study.

Although Pierre Brunet traced in often tedious detail the prolonged opposition to Newton's celestial mechanics, he says little about the early reaction in France to Newton's experiments on light and color, and what he does say is quite misleading, for he leaves the distinct impression (which perhaps he did not mean to convey) that Newton's earliest experiments must have been approved and admired, at least by 1699, the year in which Newton was elected a foreign associate (*associé étranger*) of the newly reformed Academy.[2] This was certainly not the case.[3] Many years elapsed before Newton's work in optics earned the widespread acceptance in France implied by Fontenelle's approbation. To be sure, Newton's classic first paper on light and color, printed in the *Philosophical Transactions* of the Royal Society early in 1672, was promptly read in Paris; but it was greeted at first with polite skepticism, then by hardening disapproval, and finally—for reasons I shall try to set forth—by prolonged neglect. But we may well ask: When did the tide turn in favor of Newton's doctrine? When were his classic experiments not only credited but successfully repeated in France? What was the influence of the publication in 1704 of Newton's *Opticks*, some thirty years after the appearance of the first paper, and in 1706 of the Latin version, the *Optice: sive de reflexionibus, refractionibus, inflexionibus et coloribus lucis*?

---

[2]Pierre Brunet, *L'introduction des théories de Newton en France au XVIIIᵉ siècle*, I: Avant 1738 (Paris, 1931), 8. As A. R. Hall reminds us, "Brunet was concerned almost entirely with the introduction of Newtonian celestial mechanics in France," with Newton as author of the *Principia*, not the *Opticks*. See Hall, "Newton in France," p. 235.

[3]For Newton's election in 1699, see I. Bernard Cohen, "Isaac Newton, Hans Sloane and the Académie Royale des Sciences," in I. Bernard Cohen and René Taton, eds., *Mélanges Alexandre Koyré*, 2 vols. (Paris, 1964), I, *L'aventure de la science*, 61–73. It was Newton's mathematical accomplishments, especially as demonstrated in the *Principia*, that accounted for his election as an *associé étranger*. See Chapter 3, pp. 46–47.

These questions are of more than passing interest. Newton's optical experiments, and his manner of interpreting them, announced a new methodological outlook, a subtle way of attacking physical problems that Newton, borrowing (and transforming) a term used by the early Fellows of the Royal Society, described as the "Experimental Philosophy." This new method challenged the prevailing Mechanical Philosophy (to which, of course, Newton owed an indubitable debt) and lent the final phase of the seventeenth-century scientific revolution a markedly different character. Although more than we often realize, the *Opticks* and the *Principia* speak with one voice;[4] nevertheless, it was the *Opticks*, a book that seemed to many besides Fontenelle the paramount model of experimental science, that can be said to have first breached the Cartesian defenses of the Continent, a fact of no little significance for the history of Newtonianism.

I

Much has been written about the reception accorded Newton's classic first paper, his "New Theory about Light and Colors," read on his behalf at the Royal Society of London on 6 February 1671/72. Yet there are aspects worth reviewing and some facts that need special emphasis. Of Robert Hooke's biting critique, and the controversy that long embittered the relations between the two men, there is little to add;[5] but a word or two about the response of

[4]I. Bernard Cohen in his *Franklin and Newton* (Philadelphia, 1956), chaps. 5 and 6, not only finds the *Opticks* more speculative than the *Principia*, a questionable proposition as applied to the body of the text, but holds that, in contrast to that more abstract and mathematical work, the *Opticks* exemplifies what he calls "experimental Newtonianism." But in the anonymous review of the *Commercium Epistolicum* written, or at least inspired, by Newton, we read: "The Philosophy which Mr. *Newton* in his *Principles* and *Optiques* has pursued is Experimental" (*Phil. Trans.*, 19 [1714], 222). Samuel Clarke's *Praefatio interpretis* to the Latin *Optice* (1706) also sees the two works as embodying the same approach.

[5]For the controversy between Newton and his critics there is still some value in the papers of L. Rosenberg, "La théorie des couleurs de Newton et de ses adversaires," *Isis*, 9 (1927), 44–65; and "Le premier conflit entre la théorie ondulatoire et la théorie corpusculaire de la lumière," *Isis*, 11 (1928), 111–122. Important later studies are Richard S. Westfall, "Newton and His Critics on the Nature of Colors," *Archives internationales d'histoire des sciences*, 15 (1962), 47–58; his "Newton's Reply to

scientists in France must serve as the point of departure for our inquiry.

In March 1672 Henry Oldenburg, as secretary of the Royal Society and "publisher" (we should now say "editor") of the *Philosophical Transactions,* sent to Christiaan Huygens, the chief luminary of the recently founded Royal Academy of Sciences, the number of that journal which opened with Newton's paper, specifically requesting Huygens's reaction to the new theory of color it announced.[6] But in a reply confined to what really interested him about Newton's discoveries, namely the invention of the reflecting telescope, Huygens was noncommital: "As for M. Newton's new theory of colors, it strikes me as most ingenious; but we must see if it is compatible with all the experiments."[7]

A fuller, yet unsolicited, reaction came from a young Jesuit priest, Father Ignace Pardies, a professor of mathematics at the Collège de Clermont (later the Collège Louis-le-Grand).[8] In his letter, written the same day as Huygens's reply, Father Pardies wrote that he had given careful attention to Newton's "ingenious hypothesis" and was by no means convinced of its correctness.[9] The coincidence of dates is at least suggestive, especially when we con-

Hooke and the Theory of Colors," *Isis,* 54 (1963), 82–96; and Z. Bechler, "Newton's 1672 Optical Controversies: A Study in the Grammar of Scientific Dissent," in Y. Elkana, ed., *The Interaction between Science and Philosophy* (Atlantic Highlands, N.J., 1974), pp. 115–142.

[6]*Oeuvres de Huygens,* VII (1897), 156. The letter has been reprinted in Oldenburg *Correspondence,* VIII (1971), 584–585, no. 1920.

[7]"Pour ce qui est de sa nouvelle Theorie des couleurs, elle me paroit fort ingenieuse, mais il faudra voir si elle est compatible avec toutes les experiences," *Oeuvres de Huygens,* VII, 165. All translations, unless otherwise indicated, are my own. With slight variations this passage was quoted in a letter of Oldenburg to Newton dated 9 April 1676 (O.S.) and published in Newton *Correspondence* I (1959), 135, no. 53. All letters from the Continent I leave as dated New Style; letters from England are dated Old Style, with the indication (O.S.), when dated after March 21–22.

[8]For the scant facts of Pardies's life, see Didot-Hoefer, *Nouvelle biographie générale;* Backer and Sommervogel, *Bibliothèque de la Compagnie de Jésus,* 12 vols. (1890–1911), VI, cols. 199–206, henceforth cited as Backer and Sommervogel; and Pierre Costabel's sketch in the DSB, X (1974), 314–315.

[9]For Pardies's first letter to Oldenburg, see Newton *Correspondence,* I, 130–133, and 133–134; *Newton's Papers,* pp. 79–82 (in the original Latin), and 86–89 (in an English translation from the *Philosophical Transactions Abridged,* 1809). A better translation is given in Oldenburg *Correspondence,* IX (1973), 7–10, no. 1946a.

sider Huygens's reluctance to reply in detail. Could Huygens have urged Pardies to undertake the task of examining and judging Newton's paper? This is a possibility that seems not to have been suggested, except by the present author, yet it has a certain plausibility.[10] Although Pardies was not a member of the Academy of Sciences, he and Huygens were well acquainted; and although Huygens had not yet turned his attention to the researches that culminated in his *Traité de la lumière* (1690), he knew of Pardies's interest in optics, of the Jesuit scholar's pulse theory of the nature of light, and he mentions having seen part of an unfinished treatise on light which Pardies did not live to complete.[11] Like certain other members of the Academy of Sciences, Huygens may have frequented the so-called Academy of the Abbé Bourdelot, an informal society to which Pardies belonged, and where members of the Academy could rub elbows with those scientists who, for one reason or another (such as being, like Pardies, a member of the Society of Jesus), had not been admitted to the official body.[12] Perhaps Huygens, preoccupied with more pressing matters, brought Newton's paper on light and color to the attention of the Bourdelot group, where Pardies was charged with examining it. This, of course, is conjectural; but the connection between Huygens and Pardies at this time is beyond doubt.[13]

[10]I have proposed this explanation *en passant* in Chapter 3.

[11]In a letter of Oldenburg of 24 June 1673 Huygens wrote, after criticizing one of Pardies's treatises on mechanics: "Il avoit asseurement de meilleurs choses a donner, et entre autres un petit traitè des Refractions qu'il m'a fait voir, et de belles speculations touchant le son des flutes, trompettes &c" (*Oeuvres de Huygens*, VII, 316. See also Oldenburg *Correspondence*, X (1975), 29, no. 2251.

[12]The Academy of the Abbé Bourdelot is discussed in Harcourt Brown, *Scientific Organizations in Seventeenth Century France, 1620–1680* (Baltimore, 1934), pp. 231–253. See also Paul Mouy, who writes: "Vers 1670, on y voyait de savants jésuites comme le P. Pardies, des membres de la récente Académie des Sciences, comme Gallois, Auzoult [*sic*], Pecquet, Borelli, Mariotte . . ." (*Le développement de la physique cartésienne* [Paris, 1934], p. 98). It was to the Abbé Gallois, author of the *Conversations de M. l'abbé Bourdelot* (Paris, 1672) and editor of the *Journal des sçavans*, that Huygens sent the letter describing Newton's reflecting telescope, published in that journal on 29 February 1672 and reproduced in *Oeuvres de Huygens*, VII, 134–136.

[13]See for example the letter of Pardies to Huygens, written early in July 1672, which discussed the double refraction of iceland spar. See *Oeuvres de Huygens*, VII, 193. In his *Traité de la lumiere* (1690) Huygens mentions Pardies together with

Be that as it may, Newton replied with patient civility to his French critic, explaining away certain of Father Pardies's misconceptions about the prism experiments, and protesting—as he was repeatedly forced to do henceforth—at having what he preferred to call his "theory" or his "doctrine" of color described as an "hypothesis."[14]

When Pardies raised further objections of a theoretical nature,[15] Newton again answered amiably; but took the occasion to set forth—as clearly as he was ever to do thereafter—his methodological principles and the self-imposed limitations he felt to be essential. His purpose, he insisted, was merely to determine the *properties* of light by experiment, not to explain those properties in terms of some "hypothesis" or (to translate into the modern idiom) a plausible mechanical model:

> For the best and safest method of philosophizing seems to be, first to inquire diligently into the properties of things, and establishing these properties by experiments and then to proceed more slowly to hypotheses for the explanation of them. For hypotheses should be subservient only in explaining the properties of things, but not assumed in determining them; unless so far as they may furnish experiments. For if the possibility of hypotheses is to be the test of the truth and reality of things, I see not how certainty can be obtained in

---

Hooke as having held a wave theory of light, citing "un traitté dont il me fit voir partie; et qu'il ne pût achever, estant mort peu de temps après," but in which Pardies "avoit entrepris de prouver par ces ondes les effets de la reflexion & de la refraction" (*Oeuvres de Huygens*, XIX [1937], 476). This incomplete treatise fell into the hands of another Jesuit, Father Pierre Ango, who acknowledged the use he made of it for his *Optique divisée en trois livres* (Paris, 1682), p. 14.

[14]Newton *Correspondence*, I, 140–142 and 142–144, no. 55; also *Newton's Papers*, pp. 83–85 (English, pp. 90–92).

[15]Pardies proposed that the greater length of the prismatic spectrum need not depend, as Newton thought, on the different refrangibility of the rays. It might equally well be explained by the hypothesis of Grimaldi that light is a substance moved very rapidly and subject to "diffusion" or spreading after passing through a pinhole, or by Hooke's suggestion that light is caused by undulations or pulses in a subtle medium. Newton may have learned for the first time from Pardies of the work on diffraction set forth in Grimaldi's posthumously published *Physico-mathesis de lumine* (1665). Hooke's theory was set forth in Observation IX of his *Micrographia* (1665), a work familiar to Newton.

any science; since numerous hypotheses may be devised, which shall seem to overcome new difficulties.[16]

This is a most important, as it was the earliest, formulation of Newton's scientific creed; it is one key to understanding this complex man and his ideas on method. Pardies's reply suggests that he had grasped Newton's novel point of view (for from the standpoint of the prevailing Mechanical Philosophy it was indeed novel) as neither Hooke nor Huygens seemed able to do. His doubts about the experiments themselves were removed by Newton's patient explanations, but he does not seem to have repeated any of the experiments himself.[17] Had he done so, had he been able to incorporate Newton's findings into his treatise on optics, the subsequent train of events might well have been different. But his willingness to accept Newton's results had no sequel; he published nothing further on the subject; and the following year he died at the age of thirty-seven of a fever contracted while giving the rites of the Church to prisoners in Bicêtre.

Nor, evidently, had Pardies's concession sufficed to convince Huygens, whom Oldenburg, dissatisfied with the brief comment he had elicited, continued to press for an opinion of Newton's theory of color. When Huygens at last complied, in a letter of 21 June devoted largely to discussing Newton's reflecting telescope, it was only to write rather condescendingly:

> As for his new hypothesis concerning colors, about which you wish to know my opinion, I admit that up to now it seems very plausible, and the experimentum crucis (if I fully understand it, for it is explained a bit obscurely) strongly confirms it.[18]

[16]For the English of this oft-quoted passage see *Newton's Papers*, p. 106. The Latin is given on pp. 99–103, and in Newton *Correspondence*, I, 163–168, no. 66; for an English paraphrase, see pp. 168–171.

[17]This point is made and documented in Chapter 3. Montucla (II, 621) supports this interpretation for we read: "Lorsque l'écrit de M. *Newton* vit le jour, le Père Pardies fit des objections. A la vérité, sur la réponse de M. Newton, ce Père eut la candeur de se rendre, & de témoigner qu'il étoit satisfait."

[18]Oldenburg *Correspondence*, IX, 117, gives the original French. See also *Oeuvres de Huygens*, VII, 186.

Huygens, it would seem, is echoing Pardies's generally favorable opinion; but in reality he was not wholly convinced, for he perceived difficulties of another kind. When the persistent Oldenburg pressed him once again for his opinion, he was slow to reply. At last he wrote on 27 September 1672:

> What you have published of M. Newton in one of your last numbers confirms still more his doctrine of colors. Yet the matter could well be quite otherwise, and it seems to me that he should be content that what he has put forward should pass for a very plausible hypothesis. Moreover, if it should be true that some of the rays of light are inherently [*des leur origine*] red, others blue, etc., there would still remain the great difficulty of explaining by a mechanistic physics the cause of this diversity of colors.[19]

Early in 1673 Huygens expanded upon his doubts in a letter to Oldenburg, an extract of which was printed anonymously (in translation) in the *Philosophical Transactions*.[20] In the most complete statement he was to publish about Newton's optical discoveries, Huygens did not question the validity of the experiments, nor accuse Newton of advocating a corpuscular, as opposed to an undulatory, theory of light; and he certainly did not charge Newton, even by implication (as it has been suggested Hooke may have done), with reverting to something reminiscent of Peripatetic physics and innate qualities. Huygens's objection is simply that Newton stops short of the proper goal of proposing a physical mechanism, a mechanical model, to account for his results. Until Newton can give a satisfactory mechanical explanation or hypothesis "he hath not taught us, what it is wherein consists the nature and difference of Colours, but only this accident (which certainly is very considerable) of their *different Refrangibility*."[21] In short, although Huygens

---

[19]*Oeuvres de Huygens*, VII, 228–229, and Oldenburg *Correspondence*, IX, 247–248.

[20]*Newton's Papers*, pp. 136–137, gives the portion of the letter which was published in English translation in the *Philosophical Transactions*. For the French original, see *Oeuvres de Huygens*, VII, 242–243, or Oldenburg *Correspondence*, IX, 380–381.

[21]*Newton's Papers*, p. 136. For Hooke, too, as Westfall has suggested, "the foremost issue" was the defense of the mechanical philosophy. See Westfall, "Newton and His Critics," pp. 49–51.

was prepared to lend credence to Newton's discovery that white light is composed of monochromatic rays having characteristic refrangibilities, and although he describes this discovery as "very considerable," he insisted that the main task had not been accomplished. It was necessary, above all, to account in physical terms, as befitted a mechanical philosopher, for those "original and connate properties" of the rays which give rise to the sensations of the different colors. Moreover, by asserting that *all* the hues of the spectrum are such "original and connate properties," Newton seemed to Huygens to have violated the principle of economy and to have made virtually insoluble the problem of finding an appropriate physical hypothesis. He could imagine no way of explaining by a mechanical model "so many diversities" as the infinitely many hues of the spectrum. Instead, if the great variety of hues could be thought of as produced by mixing only two kinds of colored rays (yellow and blue, for example), "it will be much more easy to find an *Hypothesis* by Motion, that may explicate these two differences."[22]

In reply, Newton could only repeat the substance of his answers to Hooke and to Pardies: that it was not his purpose to explain light or color by invoking a physical explanation, but only "to speak of *Light* in general terms, considering it abstractly," through the old geometrical device of light rays, "as something or other propagated every way in straight lines from luminous bodies, without determining what that thing is." And as to colors:

> I never intended to show wherein consists the nature and difference of colours, but onely to show that *de facto* they are originall & immutable qualities of the rays wch exhibit them, and to leave to others to

[22]Ibid. Pardies, as we noted above (n. 15), had raised a similar theoretical objection. The Honorable Roger North, who was briefly at Cambridge (Jesus College) from 1667–1669, leaving to enter the Middle Temple, and who made a reputation as lawyer and historian, wrote about 1680 that "he had much pleasure" in the theory of light and found Newton's doctrine "new and exquisitely thought." Yet he seems to have shared the doubts of Newton's Continental critics, for he wrote: "But still there wants a physical solution of this hypothesis, without which, however, it will not be admitted" (*Lives of the Norths*, 3 vols. [London, 1890], III, 63–64).

explicate by Mechanicall Hypotheses the nature & difference of those qualities.[23]

As his paper clearly shows, Huygens was well aware of Newton's position but could not accept it. With strong ties to the tradition of the Mechanical Philosophy, he failed to grasp the full significance of Newton's self-denying ordinance that gratuitous or *ad hoc* "explications" serve only to block the way to understanding, and that—as Newton later put it—"Hypotheses are not to be regarded in experimental Philosophy."[24]

Huygens maintained his position to the end. In his *Traité de la lumière* (1690) there is no mention of Newton's theory, and he remarks incidentally that the physical explanation of color is too thorny a problem to grapple with.[25] But to Leibniz, whose interest in the subject of color I shall treat later on, Huygens wrote in 1694:

> The experiments which Mr. Newton has made on the different refraction of colored rays are beautiful and interesting, but he does not explain what produces the color in these rays, a thing which I am not satisfied about either up to now.[26]

[23]Newton *Correspondence*, I, 264.

[24]*Opticks*, 2d ed. (London, 1718), p. 380. The phrase "experimental philosophy" was used by the early Fellows of the Royal Society as a portmanteau expression for their central concern with experiment. It provided the title of Henry Power's *Experimental Philosophy* (1664) and the next year Hooke used it in the preface of his *Micrographia*. As I have tried to show elsewhere, for Newton it came to have a more precise methodological meaning.

[25]*Oeuvres de Huygens*, XIX, 455. As to the explanation of colors, "personne jusqu'-icy ne peut se vanter d'avoir reussi."

[26]"Il n'explique pas ce que c'est que la couleur dans ces raions"; Huygens to Leibniz, letter of 29 May 1694, in *Oeuvres de Huygens*, X (1905), 613. Earlier he had written to Leibniz: "I have not spoken of colors in my Traité de la Lumiere, finding the matter very difficult, particularly because of the number of different ways in which colors are produced. Mr. Newton, whom I saw last summer in England, promised something on the subject, and communicated some very fine experiments from those he has made." Cited, in an English translation differing somewhat from mine, by Sir David Brewster, *Memoirs of the Life, Writings, and Discoveries of Sir Isaac Newton*, 2 vols. (Edinburgh and Boston, 1855), I, 95, n. 1. For the original of this letter of 24 August 1690, see *Oeuvres de Huygens*, IX (1901), 471.

II

In England, meanwhile, the validity and accuracy of Newton's experiments seem not to have been seriously questioned; and this was surely because Robert Hooke, in his critical attack on Newton's first paper, claimed to have satisfied himself of their correctness "by many hundreds of trials."[27] Despite his disagreements with Hooke, and their mutual distrust, Newton was not averse to citing his antagonist's testimony when doubt was cast on one of his experiments by scientists from the Continent.[28]

Little has been learned about the reception of Newton's first optical papers in English universities. It is certainly probable that at Cambridge such Newtonian disciples as William Whiston, Samuel Clarke, and Stephen Hales were aware of this early work before the publication of the *Opticks*; but the views of Roger North, described above (n. 22), must date from his London years and his reading of the *Philosophical Transactions*, not from Cambridge.

Despite the appointment of David Gregory (1659–1708), a friend and admirer of Newton, to its Savilian professorship, Oxford seems to have remained largely indifferent to, or oblivious of, Newton's theory of color. The situation in Scotland, whence Gregory came to Oxford, was vastly different. At Edinburgh, Aberdeen, St. Andrews, and Glasgow, Newton's theory of light and color was the subject of lectures and theses as early as the 1680s.[29]

[27]Thomas Birch, *History of the Royal Society,* III, 11. Reproduced in facsimile in *Newton's Papers,* pp. 110–111.

[28]*Newton's Papers,* p. 154, and Newton *Correspondence,* I, 357, no. 143.

[29]Late in 1670, James Gregory (1638–1675) at St. Andrews learned by a letter from John Collins the subject of Newton's Cambridge lectures ("Opticks proceeding where Mr Barrow left") and about Newton's telescope. On 14 March 1671/72 Collins mentioned "Mr. Newton's Experiments and Discourses about the nature of Light Colours and Refractions," Hooke's attack, and Newton's reply. See H. W. Turnbull, ed., *James Gregory Tercentenary Memorial Volume* (London, 1939), pp. 154 and 224. From Professor Eric Forbes I learned of early lectures and theses in the Scottish universities dealing with Newton's theory of light and color, first mentioned by Mary Christine King in her Edinburgh dissertation, "Philosophy and Science in the Arts Curriculum of the Scottish Universities in the 17th Century" (1974). I was able to consult some of the original documents at Edinburgh, Glasgow, and St. Andrews during my tenure of a Guggenheim Fellowship in 1978.

One matter at issue early on was Newton's simple experiment where, in the first paragraphs of his classic paper, he reported that the prismatic spectrum, in apparent defiance of the laws of optics, was not circular, but several times longer than wide. The controversy was opened in October 1674 by Francis Line (or Linus), a venerable and contentious Jesuit teaching at the English College of Liége.[30] Although loyal to the Peripatetic tradition, Line in his peculiar way was attracted to the arts of experiment, and anxious to keep in touch with recent developments. In a letter which a friend in England passed on to Henry Oldenburg, Line insisted that the lengthening of the spectrum described in Newton's first paper could be observed only when there were clouds near the sun and when the sun's rays, reflected from them, reached the prism with oblique angles of incidence; hence there was no need to assume, as Newton did, that different colored rays had different refrangibilities. When he published Line's criticism, Oldenburg added a brief explanatory reply, pointing out that Line has misconceived the conditions of the experiment: it had only been performed on clear days; the prism had been placed close to the hole in the window shutter "so that the light had no room to diverge," and the colored image produced was transverse to the axis of the prism, and not parallel to it, as Line had evidently surmised.[31]

The confusion might have been avoided if Newton, in the first instance, had described with sufficient precision and clarity the manner in which he had performed this apparently simple experiment; if he had told in his first paper exactly how to orient the prism, specified how much of the surface of the prism was to be illuminated (for the distance of the prism from the hole in the window shutter is only vaguely suggested), or done more than hint at the importance of placing the prism so that the two refractions

[30]On Francis Hall, who as a religious took the name of Linus or Line, taught Hebrew at the English College of Liége, and had some reputation as a maker of dials, see the *Dictionary of National Biography*; Backer and Sommervogel, IV, cols. 1840–1842; and C. Le Paige, "Notes pour servir à l'histoire des mathématiques dans l'ancien pays de Liége," *Bulletin de l'Institut Archéologique Liégois*, 21 (1888), 457–554, where Line is discussed on pp. 525–529.

[31]*Newton's Papers*, p. 150.

are equal. No diagram had accompanied this published description of the experiment. A man of Hooke's experimental genius could find his way through Newton's often cryptic descriptions, but pitfalls lay in the path of less perceptive men, a Pardies or a Francis Line.

In a second letter answering Oldenburg (dated 25 February 1675), Line simply clung to his assertions.[32] Newton was irritated, and ill disposed to reply, for the dispute seemed to him only about "my veracity in relating an Experiment." When, belatedly, he did reply (probably at the urging of Oldenburg, who was being importuned by Line to have his second letter published), Newton expressed doubt that Line had actually performed the experiment; probably, he wrote, he was merely relying on his "old notions." He urged that Line perform the experiment and made some suggestions on how it should be carried out successfully.[33]

Late in December 1675, Henry Oldenburg read to the Royal Society a letter from Liége which announced the death of Line, adding that his disciples proposed to repeat Newton's experiment (on the lengthened prismatic spectrum) according to the instructions of Newton's last letter. This Oldenburg took as intimating that "if the said experiment be made before the Royal Society, and be attested by them to succeed, as Mr. Newton affirmed, they would rest satisfied." It was therefore ordered "that when the sun should serve, the experiment should be made before the Society."[34]

On 12 March 1675/76, Oldenburg moved "that now the sun and season being likely to serve for making Mr. Newton's experiment called in question by Mr. Linus, an apparatus might be prepared for that purpose; Mr. Hooke said, that he had an apparatus ready to make the experiment, when the Society should call for it."[35]

When Francis Line's doubts were called to the attention of the Royal Society, Newton was pleased to report that Robert Hooke had spoken of the experiment under attack "as a thing not to be

---

[32]*Newton's Papers*, pp. 151–152, and Newton *Correspondence*, I, 334–336, no. 133.
[33]*Newton's Papers*, pp. 153–154, and Newton *Correspondence*, I, 356–358.
[34]Birch, *History of the Royal Society*, III, 271.
[35]Ibid., p. 309.

questioned." Indeed it was only to refute Line's assertions that Newton himself troubled to demonstrate the experiment in Cambridge to visitors, among them Abraham Hill, a Fellow of the Royal Society, and promised to show it at a meeting of that body,[36] which, however, he evidently failed to do. On 16 March 1675/76 it was agreed "that Mr. Newton's experiment, questioned by Mr. Linus, should be made at the next meeting, if the weather should prove favourable for it."[37] Perhaps the weather was slow to cooperate, for at the meeting of 6 April a committee was appointed "to try Mr. Newton's experiment controverted by Mr. Linus; and it was ordered that after the trial of it by that committee, it should be made before the Society."[38] The committee, consisting of Sir Jonas Moore, Dr. Croune, Abraham Hill, Nehemiah Grew, and Robert Hooke, evidently carried out the experiment successfully, with Robert Hooke doubtless in charge of the actual manipulation, for on 27 April 1676 (O.S.) the demonstration was repeated before the Royal Society "according to Mr. Newton's directions, and succeeded, as he all along had asserted it would do."[39] Soon thereafter, Henry Oldenburg informed the men of Liége of the success of the experiment.

The task of replying to Newton was left to a young student of Liége, John Gascoines. Gascoines's letter was a spirited defense of his master who, he insisted, had indeed repeated the single-prism experiment after reading Newton's paper; in fact he had "try'd it again and again" in the presence of witnesses, and probably "thrice for Mr. Newton's once, and that in a clearer and more uncloudy sky than ordinarily England doth allow." Perhaps, he remarked, the different disposition of the experiment, "the diversity of placeing the Prisme, or the bigness of the hole, or some other such circumstance," might explain the contradictory results.[40]

[36]*Newton's Papers*, p. 154, and Newton *Correspondence*, I, 357, no. 143.
[37]Birch, *History of the Royal Society*, III, 312.
[38]Ibid., p. 313.
[39]Ibid. On pp. 313–314 the method by which the experiment was carried out is described, the image being found to be "not round, as Mr. Linus contended, but oblong."
[40]Gascoines to Oldenburg, letter of 15 December 1675, in Newton *Correspondence*, I, 393–395, no. 148.

Despite the vivacity of Gascoines's letter, the young student's evident sincerity and his concern to see the matter settled by experiment somewhat mollified Newton, who set forth the precise conditions for obtaining the oblong image, and expressed the hope that when the men of Liége succeeded with the experiment they would send Oldenburg "a full and clear description" of their manner of carrying it out, illustrated perhaps by "schemes," that is, diagrams, to "make the business clearer."[41]

At Liége it was not Gascoines, who lacked the facilities or possibly the skill, to carry it out, who performed the experiment according to Newton's more specific directions. It was Anthony Lucas (1633–1693), successor to Line as professor of physics at the English College. In a letter of May 1676, Lucas—generally conceded to be one of the ablest of Newton's Continental critics—reported that he had no difficulty in verifying Newton's observation when the experiment was carried out according to Newton's instructions, although he did not find the spectrum as elongated as Newton had reported; it was not five times longer than it was broad, but only three and a half times longer, a disparity which Newton attempted in vain to account for.[42]

III

Up to this point, in the controversy between Newton and his Continental critics, only Huygens and Pardies made special mention of Newton's two-prism experiment, his *experimentum crucis*, which he described to Linus as "that which I depend on" to show

[41]Newton to Oldenburg, 15 January 1675/76, ibid., pp. 409–410, no. 15, where the whole letter is published on pp. 407–411. The portion of the letter published in the *Phil. Trans.* may be found in *Newton's Papers*, pp. 155–156.

[42]Lucas to Oldenburg, letter of 27 May 1676, in Newton *Correspondence*, II (1960), 8–12, no. 161. Reprinted from the *Phil. Trans.* in *Newton's Papers*, pp. 163–168. Lucas has long been thought to have reported his results correctly, and Newton's failure to grasp the cause of the discrepancy—that prisms, if made of different kinds of glass, can have different dispersive powers—has been held to be one of his major errors. For the contributions of Euler, Klingenstierna, and John Dolland in setting the matter straight, see Joseph Priestley, *History and Present State of Discoveries relating to Vision, Light, and Colours* (London, 1771), sect. IV, chap. I.

the different refrangibility of the solar rays.[43] Huygens, we saw, had found it "explained a bit obscurely." Pardies, for his part, had completely misunderstood Newton's original description.

In principle, as everyone knows, this experiment was elegance itself;[44] even Lucas conceded that it was an "ingenious contrivance." Yet, despite its familiarity to modern historians of science, there are some aspects that justify reviewing Newton's first description, notably the question of what, in fact, it was intended to prove. A beam of the sun's light, admitted by a hole in the window shutter of a darkened room, is passed through a prism. A board, pierced by a smaller hole, is placed "close behind the Prisme at the window." The prism, producing a vertical spectrum of colors, is so fixed that it can be rotated about its axis, allowing selected portions of the spectrum to pass through the hole of the first board. A second board, placed twelve feet away, also had a small hole for some of the incident light to pass through. Behind this second board a second prism refracted the light that had passed through the hole, and the resulting image was caught on the wall. Newton reported in rather opaque fashion that as he rotated the first prism, the second prism refracted most strongly those rays that had been most refracted by the first prism, and refracted least those that had been the least refracted by the first prism. And he concluded from this experiment that "*Light* consists of *Rays differently refrangible*" and that a spreading out of these rays, because of their different refrangibility, accounted for the lengthening of the solar image, "without any respect of a difference in their incidence," a condition deter-

---

[43]*Newton's Papers*, p. 160. Earlier, in a letter to Oldenburg of 13 November 1675 (O.S.), he had written that he wished Line would "proceed a little further to try that which I call'd the *Experimentum Crucis*." See *Newton's Papers*, p. 154, and Newton *Correspondence*, I, 357. For the origin of this expression, and Newton's debt to Hooke's *Micrographia*, see Richard S. Westfall, "The Development of Newton's Theory of Color," *Isis*, 53 (1962), 354.

[44]The famous paper in which the experiment is first described appeared in the *Phil. Trans.*, 6, No. 80, 19 February 1671/72, pp. 3075–3087, with heading: *A letter of Mr. Isaac Newton, Professor of Mathematicks in the University of Cambridge; containing his New Theory about Light and Colors [sic]: sent by the Author to the Publisher, Febr. 6. 1671/72; in order to be communicated to the R. Society.* It has often been reprinted, and is reproduced in facsimile in *Newton's Papers*, pp. 47–59, and printed in Newton *Correspondence*, I, 92–102, no. 40.

mined by the use of the holes in the two boards, and which he might well have stressed. The reason for this "extravagant disproportion" is what, as we shall see, the *experimentum crucis* was designed to demonstrate.

Lucas, of course, had the benefit of Newton's published clarifications of this experiment, but until prodded by Newton he ignored the *experimentum crucis* and sought by experiments of his own devising to show that the appearance and order of the colors "ariseth not from any intrinsecal property of refrangibility (as maintained by Mr. *Newton*) but from contingent and extrinsecal circumstances of neighbouring objects."[45] In replying to Lucas, Newton scarcely discussed his critic's experiments, and protested:

> The main thing he [Lucas] goes about to examine is, *the different refrangibility* of Light. And this I demonstrated by the *Experimentum crucis*. Now if this demonstration be good, there needs no further examination of the thing; if not good, the fault of it is to be shewn: for the only way to examine a demonstrated proposition is, to examine the demonstration. Let that Experiment therefore be examined in the first place, and that which it proves be acknowledged, and then if Mr. *Lucas* want my assistance to unfold the difficulties which he fancies to be in the Experiments he has propounded, he shall freely have it.[46]

On October 23 Lucas wrote again to Oldenburg, protesting, without the addition of confirmatory experiments, that the *experimentum crucis* was no demonstration of Newton's theory. He insisted once again that these supplementary experiments proved "that this inequality of refraction at an equall incidence seldome accompanys different colours." Indeed the *experimentum crucis* struck him as being "quite contrary to Mr. Newtons assertion." His repetition of this experiment—the first, I believe, to be carried out on the Continent—is described by Lucas as follows:

[45]*Newton's Papers*, p. 168, and Newton *Correspondence*, II, 11, no. 161.

[46]*Newton's Papers*, p. 174, and Newton *Correspondence*, II, 79–80, no. 173. Here too Newton makes the significant remark that it will "conduce to his [Lucas's] more speedy and full satisfaction if he a little change ye method wch he has propounded, & instead of a multitude of things try only the *Experimentum Crucis*. For it is not number of Expts, but weight to be regarded; & where one will do, what need of many?" A similar dictum was set forth earlier by Isaac Barrow, Newton's friend and, to a degree, his mentor, in his *Mathematical Lectures*.

In order then to this experiment I placed the first prisme as in my former letter [i.e., two inches from the hole]: the 2d, distant from the window about 12 foot: the distance of the paper (wheron the image was cast) from the second prisme about 3 foot: the two boards at the prismes were placed exactly according to Mr. Newtons directions.

Then followed the result of the experiment:

Upon casting the violet rayes on ye second prisme, I constantly indeed found a considerably greater refraction than when the red ones were cast theron: yet withal (notwithstanding all care in excludeing extraneous light) I as constantly found thes violet rayes accompanyed by a considerable quantity of red ones, upon ye paper behind the second prisme.[47]

This being so, Lucas reasoned, red rays, although refracted by the first prism, remained "still interlaced with the violet ones," and therefore the "red rays suffered the same refraction in the first prisme at the window, as the violet ones did."[48]

Newton's reply is extremely interesting. In his letter of 5 March 1677/78, he wrote:

Having thus run over your letters let me add yt all your Objections run upon two general mistakes. You mistake both ye question of different refrangibilities & ye notion of primary colours.

And, he continued, the question of different refrangibility which the *experimentum crucis* is to decide

is not, as I sayd whither [*sic*] rays differently coloured are differently refrangible, but only whether some rays be more frangible yn others. What ye colours of ye rays differently refrangible are, or whither [*sic*] they have any appropriate ones belongs to an after enquiry.[49]

---

[47]Lucas to Oldenburg, dated Liége, 23 October 1676, in Newton *Correspondence,* II, 104–108, no. 185, where the experiment is described on pp. 105–106. These later exchanges between Lucas and Newton were first printed in the Newton *Correspondence* and are not included in *Newton's Papers.*

[48]Newton *Correspondence,* II, 106.

[49]Newton to Lucas, 4 March 1677/78, ibid., pp. 257–258.

Newton then reminds Lucas that in his first paper on light and colors "I make no mention of colours while I am prouving different refrangibility by ye *Experimentum Crucis,* but after yt begin to tel you yt ye rays wch differ in refrangibility differ also in colour, *reserving this to be proved by other Experiments.*"[50]

Now this may seem merely an adroit defensive maneuver, but a careful scrutiny of Newton's first paper supports his surprising statement. In that paper the awkwardness of his description of the *experimentum crucis* seems to have resulted directly from avoiding any mention of color. Newton did *not* say that the violet rays (most refracted in the first prism) are most refracted in the second prism, or that the red rays are least refracted in both. Indeed it is only *after* his brief discussion of the *experimentum crucis,* followed by a digression about the reflecting telescope (presented as a result of his discovery of the different refrangibility of rays), that he turns to "another more notable difformity" in the rays of light "wherein the *Origin of Colours* is unfolded." At this point he sets forth what he calls his theory or doctrine of color, enunciated in thirteen propositions that he evidently believed must be tested by special experiments. For our purposes the most important of the propositions are the first three: (1) just as rays of light differ in their degrees of refrangibility "so they also differ in their disposition to exhibit this or that particular colour"; (2) with the same degree of refrangibility there is always associated the same color and vice versa; this "Analogy 'twixt colours, and refrangibility, is very precise and strict"; and (3) "the species of colour, and degree of Refrangibility proper to any particular sort of Rays, is not mutable by Refraction, nor by Reflection from natural bodies, nor by any other cause, that I could yet observe."[51]

In what follows Newton makes passing reference to experiments made to confirm certain of these propositions, but the only one he describes in detail, and with a diagram, has little to do with the first three propositions; it is the experiment with a prism and a biconvex lens which shows that, whereas the prism separates the white light

[50]Ibid., p. 258. The concluding emphasis is mine.
[51]*Newton's Papers,* p. 53.

into the various colored rays of the spectrum, the lens, in focusing the rays, reproduces or synthesizes the whiteness.

Newton seems well aware that nowhere has he described a conclusive experiment to support the main thrust of this doctrine: the inseparable relation between refrangibility and color, and the impossibility of further altering monochromatic rays by refraction or other means. So at the close of his famous paper he writes:

> If you proceed further to try the impossibility of changing any uncompounded colour . . . 'tis requisite that the Room be made very dark, least [sic] any scattering light, mixing with the colour, disturb and allay it, and render it compound. . . . 'Tis also requisite, that there be a perfecter separation of the Colours, than, after the manner above described, can be made by the Refraction of one single Prisme, and how to make such further separations, will scarce be difficult to them, that consider the discovered laws of Refractions. But if tryal shall be made with colours not throughly separated, there must be allowed changes proportionable to the mixture.[52]

Clearly, Newton—although confident in his doctrine—was aware that he had not described in sufficient detail how truly "monochromatic" rays could be produced, or how to prove that color and refrangibility are indissolubly linked. In the summer of 1672, Newton wrote a brief letter to Oldenburg listing a series of eight "Quaeries" to be investigated by further experiment. The third of these asks: "Whether rays, which are endued with particular degrees of refrangibility, when they are by any means separated, have particular colours constantly belonging to them?" The fourth reads: "Whether the colour of any sort of rays apart [i.e., separated] may be changed by refraction?"[53]

And Newton concluded: "To determine by Experiments these and such like *Quaere's* [sic] which involve the propounded Theory, seems the most proper and direct way to a conclusion. . . . For if the

---

[52]Ibid., p. 59. This important caveat is later elaborated in the *Opticks* (1704), and methods of separating the rays described, in Book I, Part I, Prop. IV, Prob. I, pp. 45–51.

[53]Ibid., pp. 93–94. These "Quaeries" were printed in both Latin and English in the *Phil. Trans.*, 7 (1672), no. 85, 5004–5007.

Experiments, which I urge, be defective, it cannot be difficult to show the defects; but if valid, then by proving the Theory they must render all Objections invalid."[54]

It should be clear, from what has just been presented—Newton's protest to Lucas, the very structure of the famous first paper on light and colors, the "Quaeries" of July 1672—that Newton made a distinction betwen the *factual discovery* of the *experimentum crucis* (that light is composed of rays differently refrangible) and the theory or doctrine of color. Yet Lucas, and others of Newton's contemporaries, including France's leading experimental physicist, Edme Mariotte (ca. 1620–1684), had reason to believe that the *experimentum crucis,* as Newton first described it, alone would suffice to prove or disprove Newton's doctrine.

Born in Burgundy, and a founding member of the Royal Academy of Sciences (1666), Mariotte was the first in France to carry out experiments on the physiology of plants; he discovered the blind spot in the human eye; and he independently hit upon the relation of the pressure and volume of a mass of air, and more clearly than Boyle recognized it as a "sure rule or law of nature." In 1681 Mariotte published a little tract, *De la nature des couleurs,* which henceforth was taken in France to be the decisive refutation of Newton's experiments. Shortly before this date, precisely when I have not ascertained (though perhaps early in 1679, for in that year he presented his results to the Academy), Mariotte undertook an experimental verification of Newton's results. Some of his experiments, so he tells us, accorded well enough with the "hypothèse nouvelle" of the learned Mr. Newton, but others seemed to refute it. Chief among the latter was the *experimentum crucis,* which Mariotte believed he had performed after Newton's fashion, but which did not succeed as Newton said it should! Mariotte passed a narrow beam of sunlight through a prism and caught the spectrum on a piece of white cardboard. When only the violet rays were then allowed to pass through a slit (*fente*) in the cardboard and were again refracted by a second prism, Mariotte

[54]Ibid.

found the emerging violet rays tinged with red and yellow. Obviously, he concluded, they had been altered or "modified" by their passage through the second prism, and he remarks:

> From this experiment, it is evident that a given portion of light receives different colors as a result of different modifications, and that the ingenious hypothesis of Mr. Newton should not be admitted.[55]

To nearly all Mariotte's followers this seemed conclusive. His reputation as one of France's outstanding experimenters lent great weight to his pronouncements, and we have ample testimony that more than any other factor his failure to confirm the *experimentum crucis* long delayed the acceptance of Newton's theory of color.[56] After the publication of Mariotte's tract, Newton's optical discoveries were generally ignored on the Continent—Leibniz, as we shall see, was a unique and persistent exception—for some thirty years. Except for the brief notice, cited earlier, of Mariotte's work on color, no significant contribution on this subject was made at the Academy of Sciences until 1699, when the distinguished philos-

[55]E. Mariotte, *De la nature des couleurs* (Paris, 1681), p. 211; also *Oeuvres de Mariotte*, 2 vols. (Leiden, 1717), I, 227–228. In the "édition nouvelle" of 1740 (two vols.-in-one, consecutively paginated), the work on colors is found on pp. 196–320. Here Mariotte's discussion of Newton's theory in the light of his experiments is on pp. 226–228. Mariotte presented his paper to the Academy of Sciences in 1679. See J. B. Du Hamel in *Regiae scientiarum academiae historia*, 2d ed. (Paris, 1701), p. 184, where we read: "Tractatum suum de coloribus legit D. Mariotte, quem postea in publicum emisit." The first edition of Du Hamel's history was published in 1698. A longer account is given in *Hist. Acad. Sci.*, I, 1666–1686 (Paris, 1733), 291–303.

[56]Voltaire wrote: "On s'est d'abord révolté contre le fait [the different refrangibility of the differently colored rays], & on l'a nié long-tems, parce que M. Mariote [*sic*] avoit manqué en France les expériences de Neuton," *Elémens de la philosophie de Neuton* (Amsterdam, 1738), p. 121. See also Montucla II, 622, and Pierre Coste, who in the preface to his translation of Newton's *Opticks* (Amsterdam, 1720) described Mariotte as having repeated Newton's experiments "d'une manière imparfaite." Condorcet later wrote in his *éloge* of Mariotte: "Avouons, cependant, que, lorsqu'il essaya de répéter les expériences de Newton sur la lumière, il les manqua, & qu'il établit, d'après les résultats des siennes, un systême tout différent de la théorie de Newton. Dans cette seule occasion, la sagesse de son esprit & sa dextérité dans les expériences semblent l'avoir abandonné à-la-fois." See *Eloges des académiciens de l'Académie royale des sciences morts depuis 1666, jusqu'en 1699* (Paris, 1773), p. 60. See also A. Saverien, *Histoire des progrès de l'esprit human* (Paris, 1766), p. 267, and Priestley, *History of Vision, Light, and Colours* (London, 1772), p. 350.

opher, Father Nicolas Malebranche, recently elected in *académi-cien honoraire* of that body, presented his "Réflexions sur la lumière et les couleurs." Ignorant of Newton's experiments and his theory of the composition of solar light, or perhaps merely persuaded, as were other Frenchmen, of their inaccuracy, Malebranche made no mention of Newton. Instead, he set forth his own theory, about which he had previously dropped hints in his *Recherche de la vérité*, that departed from the widely accepted color theory of Descartes, his master in so many other respects. Notably, he undertook to do precisely what many years earlier Huygens had criticized Newton for not attempting: that is, to supply a plausible hypothesis, or mechanical model, to account for the origin of the "colorific" prop-erties of the different rays.[57] We shall soon return to the influence of Malebranche—for it is centrally important to our narrative—when the publication of Newton's *Opticks* in 1704, and of Samuel Clarke's Latin version of the work two years later, aroused interest once again in Newton's doctrine of color.

IV

Although written in English, a language little read in France at the time, Newton's *Opticks* attracted attention not only for the text itself but also for the two Latin mathematical tracts appended to it, one on curves of the third degree, the other Newton's first pub-lished paper dealing with the calculus, his *De quadratura curvarum*. This English first edition of the *Opticks* soon reached France. One such copy was sent by Sir Hans Sloane, one of the two secretaries of the Royal Society, to Etienne François Geoffroy (1672–1731), a leading chemist and physician of the Academy of Sciences. Geof-froy wrote to Hans Sloane on 20 January 1706 to report that he had received everything Sloane had sent him since the beginning of the war, that is, the volumes of the *Philosophical Transactions* for 1702 and 1703 and "l'excellent traitté d'Optique de Mr. Newton

[57]Pierre Duhem, "L'optique de Malebranche," *Revue de métaphysique et de morale*, 23 (1916), 37–91.

Plate 1. Etienne François Geoffroy (1672–1731). Courtesy of the Bibliothèque Nationale, Paris.

que le père de fontenay ma remis entre les mains de votre part."[58]

Sloane had been a faithful correspondent of Joseph Pitton de Tournefort, a famous botanist with whom he had studied in Paris. In February 1698 Tournefort wrote Sloane introducing "M. Geoffroy qui est un parfait honnete homme tres curieux, habile dans l'histoire naturelle et avec qui je vis depuis fort longtemps en parfaite amitié."[59] Later that year Geoffroy traveled to England as the personal physician of the Comte de Tallard when that distinguished soldier and diplomat was sent, at the conclusion of the Treaty of Ryswick, to serve as *ambassadeur extraordinaire* to work out the final territorial settlements.[60] On that occasion, Geoffroy made several appearances at the Royal Society, the earliest being on April 13 (O.S.) if the record is to be trusted, during which he made some chemical demonstrations.[61] On June 29 he was "proposed by Dr. Sloane a Candidate to be elected a Member of the Society,"[62] and was duly elected shortly after. The following year the Academy of Sciences in Paris reciprocated by naming Sir Hans Sloane as a *correspondant*. We may conclude, I think, that the presentation of Newton's *Opticks* to Geoffroy was an official courtesy of one scientific society to the other. This copy was not in any direct sense a presentation copy from Newton, but one of a certain number sent by the printer to the Royal Society, at the author's request, for distribution as the Secretaries saw fit.

At all events Geoffroy made good use of his. He not only read it with care, but prepared an elaborate précis in French which he

[58]British Library, Sloane MSS 4040, fol. 114. The war, now known as the War of the Spanish Succession, began early in May 1702. The delay before Geoffroy received Sloane's gifts is easier to understand than the successful exchange of scientific information in time of war. The "père de fontenay," whom I have not identified, may have been a Jesuit Father, Pierre-Claude Fontenai (1663–1742), remembered as a church historian. Could he have been a secret agent of the French crown? For Geoffroy see the recent sketch by W. A. Smeaton in DSB, V (1972), 353–354. Geoffroy's copy of the *Opticks* (1704) is now in the History of Science Collections of Cornell's Olin Research Library.

[59]Cited by Jean Jacquot, "Sir Hans Sloane and French Men of Science," *Notes and Records of the Royal Society of London*, 9 (1952), 87.

[60]Jean Jacquot, *Le naturaliste Sir Hans Sloane (1660–1753) et les échanges scientifiques entre la France et l'Angleterre* (Paris, 1953), p. 15.

[61]Journal Book, Royal Society of London, III, 67, entry for 13 April 1698.

[62]Ibid., III, 75, entry for 29 June 1698.

read to the members of the Academy in a series of sessions, not always consecutive, between 7 August 1706 and 1 June 1707. His audience, it is worth noting, included such important figures of our story as Father Malebranche, Jean Truchet (Père Sébastien), the physicist Pierre Varignon, and the *secrétaire perpétuel*, Fontenelle. The first entry of 7 August 1706 reads: "M. Geoffroy a commencé à lire une traduction qu'il a faite en forme d'Extrait d'un Traité Anglois de M. Neuton sur l'optique."[63]

Clearly, then, it was the English *Opticks* that Geoffroy described to his colleagues, not—despite the dates of his presentation, explained by the delayed receipt of the book—the Latin *Optice* of 1706. Indeed there can be little doubt that the very document he read before the academicians can be seen at the Bibliothèque de l'Arsenal (Paris). It is an anonymous manuscript of 127 pages, written in an eighteenth-century secretarial hand and bound handsomely in marbled calf. Entitled *Optique, ou traité des réflexions, réfractions, inflexions et couleurs de la lumière, par M. Neuton. 1704*, it is not a word-for-word translation but a précis composed in the third person.[64]

This English *Opticks* was ignored by the editor of the *Journal des sçavans* which, however, was to print a detailed résumé of the Latin *Optice* of 1706. By contrast, the *Acta eruditorum*, the learned journal we associate with Leibniz, was more alert: the issue of January 1705 contains a review of the mathematical papers appended to the 1704 *Opticks*; and a year later, in February 1706, it printed a detailed account of the *Opticks* proper, in which Newton is described as setting forth his theory of light and color, not by means of hypotheses, "sed per rationes & experimenta proprietates luminis explicans."[65] This was a clear and dispassionate summary, not only of Newton's dispersion theory of color, but also of the principal

[63]*Procès-verbaux*, 1706, fol. 321.

[64]Bibliothèque de l'Arsenal, 2883 (97. S.A.F.). I. Bernard Cohen was the first historian of science to call attention to this manuscript, and to suggest that it "may very well be the text used by Geoffroy for his report to the Academy of Sciences." See his "Isaac Newton, Hans Sloane and the Académie Royale des Sciences," in *Mélanges Alexandre Koyré*, I, pp. 102–116. A brief description was given in Henry Martin, *Catalogue des manuscrits de la Bibliothèque de l'Arsenal*, III (1883).

[65]*Acta eruditorum* (February 1706), pp. 59–64.

experiments by which he was led to it. We shall probably not err in attributing the review to Leibniz himself, for Leibniz—as we shall soon see—was keenly aware of the importance of Newton's putative discoveries, and anxious to determine whether or not his theory of color should be accepted.

Sixteen years were to elapse, however, before a French translation of the *Opticks* was published. To be sure, the accessibility of a Latin version to which French scientists could turn explains in part why a French translation was not essential, and why its appearance was so long delayed. Yet I believe the more cogent reason was that there was as yet no convincing confirmation of Newton's experiments; Mariotte's refutation, which carried great weight, was reprinted at least twice between 1681 and 1720, the year when a French translation of the *Opticks* finally appeared. It is perhaps significant that the problem of color, with no reference to Newton's theory, was the subject of only a single communication to the Academy of Sciences—that of Father Malebranche—during the period just treated. There are, however, other sorts of evidence that Newton's book had excited some interest and made converts in the world of French science. To these we must now turn.

Some information can be gleaned from two small pieces that Father Antoine Laval appended to his *Voyage de la Louisiane* (1728).[66] A reputable Jesuit astronomer, who now and again sent communications to the Academy of Sciences in Paris, Laval for most of his career taught mathematics and hydrography to naval officers first at Marseilles, where he had an observatory in the Jesuit house of Sainte Croix, which he used from 1706 to 1718, and

---

[66]See Chapter 4. Laval's book is entitled *Voyage de la Louisiane, fait par ordre du Roy en l'année mil sept cent vingt: Dans lequel sont traitées diverses matieres de Physique, Astronomie, Géographie & Marine . . . Et des Reflexions sur quelques points du Sistème de M. Newton*. Par le P. Laval de la Compagnie de Jésus (Paris, 1728). The *Voyage* itself, after the front matter, occupies pp. 1–304. This is followed by "Observations sur la refraction faites à Marseilles" separately paginated pp. 1–96; followed by "Recueil de Divers Voyages," describing, pp. 1–151, astronomical observations in the south of France. The Newtonian papers are at the end of the volume, pp. 153–191; they include Laval's short "Reflexions sur quelques endroits du Traité d'Optique de Monsieur le Chevalier Nevvton," pp. 184–191. The best account of Laval is in Backer and Sommervogel, IV, cols. 1570–1575.

then at Toulon, where he died in the year his book appeared. Ten years before, in 1718, there had taken place an exchange of letters between two naval officers of his acquaintance, one of whom, at this early date, was a convinced Newtonian, anxious to convince his colleague that Newton, in his nautical metaphor, "avoit *coulé à fond* Monsieur Descartes." By mutual agreement the two officers sent to Father Laval, to elicit his opinion, the letters they had exchanged. This caused Laval to set down some reflections of his own on Newton's new system of the world, which he dispatched to one of his officer friends and later appended to his *Voyage.* From this we learn that the Newtonian officer, whom Laval simply calls "M. le Chevalier," was familiar with Newton's *Principia,* Edmond Halley's work on cometary theory, and Samuel Clarke's Latin version (1697) of Jacques Rohault's *Traité de physique,* embellished with Newtonian footnotes. More immediately relevant to our subject, he had expounded at length to his colleague, as Laval tells us, the subject of Newton's theory of color. The source of the officer's information, so a specific page reference discloses, was the Latin *Optice* of 1706. What is especially interesting is that Laval himself, despite a considerable familiarity with the *Principia* of Newton, in 1719 had not read the *Opticks*: "Je ne sçavois point pour lors que l'optique de Monsieur Newton eut été traduite de l'Anglois; ainsi pour mon malheur je ne l'avois pas lûë."[67] His familiarity with that work, which he discusses in a short paper at the end of the *Voyage de la Louisiane,* had to await the return of the expedition aboard the *Toulouse,* a voyage that lasted from the end of March to the middle of November 1720. Laval's "Réflexions sur quelques endroits du Traité d'Optique de Monsieur le Chevalier Newton," were written for his Newtonian friend and were based, so he tells us, on his perusal of the second edition (1722) of Coste's French translation of the *Opticks,*[68] of which more later.

A brief summary of Laval's "Réflexions" must suffice. He has

---

[67]*Voyage de la Louisiane,* "Reflexions," p. 153.

[68]Laval, believing that his correspondent, the mysterious M. le Chevalier, had used the first French edition of the *Opticks,* sought in vain to locate a copy in 1719, a year before the Amsterdam edition was actually published.

much praise for "cet illustre et sçavant Auteur." The experiments and inferences of the two parts of Book I "sont quelque chose d'admirable";[69] he is flattering about the first two parts of Book II ("On ne sçaurait trop méditer sur d'aussi belles & aussi originales découvertes");[70] but he is less happy with Newton's attempts to explain why some bodies reflect certain colors more strongly than others, or with the proposition that reflection does not involve the actual contact of rays of light with the surface of bodies. As a Cartesian, Laval scouts any thought of *actio ad distans*, opposing the concept of attraction as it appears in the Queries. Moreover, as a good Catholic, he is particularly offended by the conclusion of Query 31 where Newton asserts that if natural philosophy be properly pursued, it can improve moral philosophy beyond the four cardinal virtues to which even the Heathen, though blinded by the worship of false Gods, had risen. Does he expect, the Jesuit Father wondered, that natural philosophy could imbue men with the theological virtues ("les vertus Théologales"), that is, with faith, hope, and charity? Only the Christian church can instill these virtues "que la raison humaine n'y sçauroit parvenir par ses propres forces."

That a scientist of Laval's competence, although familiar in some degree with the *Principia*, should have been ignorant of Newton's *Opticks* before 1722, is at least suggestive.[71] Did this ignorance generally prevail in France? Or were there others who, in the years between 1706 and 1720, shared the enthusiasm of our unidentified "M. le Chevalier" for Newton's theory of light and color?

In any case the evidence we have extracted from Laval's book already suggests that the publication of the *Opticks* served as the efficient cause in overcoming those doubts about Newton's experiments which Mariotte had sown abroad. But what of Geoffroy's

[69]Book I of the *Opticks* is chiefly devoted to proving the heterogeneity of white light and the relation of rays of different colors to their refrangibility.
[70]Concerning the reflections, refractions, and colors of thin transparent bodies.
[71]Laval's ignorance is hard to understand, for a review of the Latin *Optice* of 1706 appeared in the *Journal des sçavans* for October 1707, pp. 137–149. See Jacqueline de la Harpe, *Le Journal des savants et l'Angleterre,* University of California Publications in Modern Philology, 20, no. 6 (1941), p. 322.

exposition of Newton's discoveries before the Academy of Sciences? Did it have a demonstrable effect? The answer, I believe, must be an unconditional yes. As Pierre Duhem has clearly shown, the venerable Father Malebranche was the earliest Frenchman of real eminence and authority to take seriously Newton's work on light and color.[72] Duhem was, however, unaware that Malebranche, a faithful attendant at the meetings of the Academy of Sciences,[73] had heard Geoffroy announce the appearance of Newton's *Opticks* and over a number of sessions summarize its contents.

Before we take up Malebranche's conversion, we should say a word or two about the theory he set forth in his paper "Réflexions sur la lumière et les couleurs" of 1699.[74] That he owed a considerable debt to Huygens as well as to Descartes is quite apparent; but his doctrine of color is uniquely his own and might be described (although such comparisons are hazardous) as the earliest anticipation of an undulatory theory *of color*. Light, to Malebranche as to Pardies, Hooke, and Huygens, is nothing but a pulse or vibratory motion of the aether or subtle matter; and for Malebranche, colors (or at least certain of them) are the result of the differing frequencies of these pulses, just as pitch is determined by the frequency of sound waves in air produced by a vibrating source. Malebranche's subtle matter, like the "second matter" of Descartes, is particulate; but unlike the little hard spheres of the Cartesian aether, the particles imagined by Malebranche are compressed fluid masses, tiny vortices that, because they are fluid, can transmit vibrations simultaneously in all directions. Not all the hues of the spectrum are

[72]See his classic study, "L'optique de Malebranche," *Revue de métaphysique et de morale*, 23 (1916), 37–91. Consult also the important paper of P. Mouy, "Malebranche et Newton," *Revue de métaphysique et de morale*, 45 (1938), 411–435.

[73]André Robinet, "La vocation académicienne de Malebranche," *Revue d'histoire des sciences*, 12 (1959), 3–4. Malebranche was present at five of the Academy's sessions in which Geoffroy read his French version of Newton's *Opticks*: August 7, 11, 14, of 1706, and January 15 and and 26, of 1707. See Correspondance, actes et documents, 1690–1715, ed. Robinet, in *Oeuvres de Malebranche*, XIX (Paris, 1961), 732 and 749.

[74]*Mem. Acad. Sci.*, année 1699 (Paris, 1702), pp. 22–36. A good summary is given in Duhem, pp. 78–84.

produced by this mechanism, only the primaries (*primitives*) red, yellow, and blue. From these, the other hues arise by mixture. The sensation of black results when no vibrations at all are transmitted or reflected to the retina; white, when the original vibrations set up by luminous source are unaltered by reflection, refraction, or the conditions of transmission. This is manifestly what has been called a "modification" theory, since the primary hues are produced when white light is in some way altered in frequency by some medium encountered in its path. Yet the model has the interesting feature that it could be, and subsequently was (by Malebranche himself), rather drastically changed to accommodate Newton's experimental results and the notion of the prior existence in a beam of white light of an infinite range of qualities producing the infinite range of colors, for all one had to do was to imagine a corresponding infinitude of possible frequencies.

Paul Mouy made clear that Malebranche discovered Newton's work on color between the year 1700, when there appeared the fifth edition of the *Recherche de la vérité*, and 1712 when the sixth edition was published.[75] As early as the third edition of his famous work, Malebranche had adopted the practice of adding appendices that he called elucidations (*éclaircissements*). In the fifth edition of 1700 Malebranche reprinted, virtually unaltered as a new "elucidation," his 1699 paper on light and color. But in the edition of 1712, the sixth, this "XVIe éclaircissement" is markedly changed.[76] The "primitive" colors of the earlier paper are now spoken of, in Newton's terminology, as "simple" and "homogeneous"; and white light ("the most composite of all") is described, according to his own adaptation of Newton, as "composed of an assemblage of different vibrations" of the subtle matter. Each color, Malebranche argued,

[75]Mouy, p. 419. Duhem makes the same point, but less strongly, and confuses the successive editions of *De la recherche de la vérité*, calling that which appeared in 1700 the third edition, and that of 1712 the fourth.

[76]See *De la recherche de la vérité*, ed. F. Bouillier, 2 vols. (Paris, 1880), II, 481–530; or *Oeuvres de Malebranche* (De la recherche de la vérité, Eclaircissements), III (Paris, 1964), 255–305. For the changes Malebranche made in treating color, see André Robinet, *Malebranche de l'Académie des sciences* (Paris, 1970), pp. 299–306 and 317–323.

has its own characteristic refraction; to be convinced of this, one need only consult the experiments in the "excellent work of M. Newton." And to the original *éclaircissement* so emended Malebranche further added several pages of a "Proof of the Supposition I have made," in the course of which he again cited Newton's experiments.

As Duhem was the first to point out, it can only have been the appearance of Newton's *Opticks* (in English in 1704 and in Latin in 1706) that accounts for Malebranche's conversion to Newton's doctrine,[77] as indeed his own words seem to indicate. In any case, there is supporting evidence. In September 1707, J. Lelong, Leibniz's faithful correspondent in France,[78] wrote a letter in which he remarked to Leibniz that the Reverend Father Malebranche has been for some time in the country, where he is attempting to confirm (*où il vérifie*) the optical experiments of Mr. Newton. In a letter of 23 October, the emphasis is different. Malebranche, he writes, "has taken with him a work of Mr. Newton, printed in London, 1704, *De Quadratis* [*sic*] *Curvarum*, in which this author pushes the integral calculus farther than everything that has been printed up to now."[79] A variant of this letter mentions that this little treatise is appended to Newton's "treatise on colors," that is to say, the *Opticks*.[80] In his October letter Lelong is definite that Father Malebranche's interest was primarily in the mathematical tracts, which is indeed likely if André Robinet is correct that Malebranche was *en villégiature* with his mathematician friend Rémond de Montmort.

---

[77]Duhem, p. 85. Duhem speaks only of the Latin *Optice*, as does Robinet.

[78]Jacques Lelong (1665–1721), an Oratorian priest was, as librarian of the Oratoire de Paris, in daily contact with Malebranche.

[79]Malebranche, *Oeuvres de Malebranche* (Correspondance, Actes et Documents, 1690–1715), XIX, 768. This letter is cited by Mouy (p. 419) from the Abbé Blampignon's *Etude sur Malebranche d'après les documents manuscrits, suivie d'une correspondance inédite* (Paris, 1862), pp. 138–139. The "Correspondance inédite de Malebranche" is separately paginated.

[80]Blampignon's variant (pp. 138–139) reads: "Ce Père est à la campagne avec un de ses amis. Ils ont emporté le nouveau livre de M. Newton ou son traité des couleurs, à la fin duquel il y a un petit traité *De quadraturis* [*sic*] *curvarum*, où il pousse le calcul intégral plus loin que tout ce qu'on avoit vu d'imprimé jusqu'à présent." Blampignon's error of assigning Lelong's letter to the year 1704 is copied by Mouy (p. 419).

That Malebranche did not read English was no obstacle if his concern was primarily mathematical, so it is likely that what he took to the country was the English *Opticks* of 1704. But he soon studied the main work on light and color in the Latin *Optice*, which had appeared the year before Lelong wrote and which is the edition Malebranche subsequently cited.[81]

There is no confirmation of the statement in Lelong's first letter that Malebranche attempted to repeat the experiments of the *Opticks*. But this was the interpretation of Leibniz when he wrote not long after, on 13 December 1707, to Lelong:

> I should be curious to learn what the Reverend Father Malebranche may have observed about colors. The subject is important. There is an experiment of M. Newton which M. Mariotte has challenged, and which should be examined above all. For M. Newton claims that one can separate the colored rays from one another, so that after this separation refraction does not make them change color any further.[82]

It seems to have been in the same year that Malebranche wrote to a friend:

> Although M. Newton is not a physicist, his book is very curious and very useful to those who have the right principles in physics, and he is moreover an excellent mathematician. Everything I believe concerning the properties of light fits all his experiments [*s'ajuste à toutes ses expériences*].[83]

[81]Mouy (p. 420) writes: "Il semble bien que Malebranche ne lût pas l'anglais, et, en tout cas, lorsqu'il citera l'*Optique* de Newton, ce sera d'après l'édition latine de 1706, non d'après l'édition anglaise de 1704." For Robinet's conjecture that Rémond de Montmort was Malebranche's host in the summer of 1707, see his *Malebranche*, p. 300.

[82]*Oeuvres de Malebranche*, XIX, 769; André Robinet, *Malebranche et Leibniz, relations personnelles* (Paris, 1955), pp. 355–356; Blampignon, *Etude sur Malebranche*, p. 137. In each source the printed text reads: "Car M. Newton pretend qu'on ne [*sic*] peut separer les rayons colorés les uns des autres, en sorte qu'après cette separation, la refraction ne les fait plus changer de couleur." This must have been a slip of the pen or an error in transcription. As printed it does not represent Newton's views, which Leibniz understood quite well, and the negative in the first phrase conflicts with the last.

[83]*Oeuvres de Malebranche*, XIX, 771–772. This letter, not dated by Blampignon, (*Correspondance inédite*, p. 25) is tentatively assigned by Robinet to the year 1707.

That members of Malebranche's scientific circle did more than simply read Newton's book, and may have undertaken to verify the dispersion experiments, is implied in the following passage in a letter, dated 9 April 1708, from Leibniz to Lelong:

> I fear that M. de La Hire may have tried the experiments of M. Newton on colors with some preconceptions, and may not have used all the care that could be given to them. For, since M. Newton has worked at them for so many years, and since one cannot doubt his ability, it is not credible that he has recounted imaginary experiments. So I should wish that persons with all the necessary leisure, and who are willing to apply themselves sufficiently (which one should not ask of persons of the age and distinction of the Reverend Father Malebranche and M. de La Hire) might be entrusted with this inquiry: this is what I have written to M. l'Abbé Bignon.[84]

From this it would appear that Philippe de La Hire (1640–1718), the well-known mathematician and the friend and Academic colleague of Mariotte, tried, but without success, to repeat Newton's experiments. But that Malebranche also attempted an experimental confirmation is doubtful.[85] There is no evidence that he did; it is more likely that he simply took Newton's account at face value. As Duhem put it:

> The reading of the *Opticks* of Newton inspired him [Malebranche] with the greatest confidence in the admirable experiments that this book reported; he hastened to retouch his theory of colors, so that it would accord fully with the truths demonstrated by the great English physicist.[86]

---

[84]Blampignon, *Correspondance inédite*, p. 137; and Robinet, *Malebranche et Leibniz*, pp. 355–356. The Abbé Bignon was the "*président*" of the Academy of Sciences, a key figure in its renovation in 1699. An Oratorian, like Malebranche, he died in 1743 and his *éloge* was written by Dortous de Mairan.

[85]The word "vérifie" in Lelong's letter of 4 September 1707 may suggest an attempt to "verify" Newton's results experimentally; but this is doubtful in view of Malebranche's age, his theoretical and bookish approach to physical science, and because he is not otherwise known to have performed experiments, although accounts of them interested him.

[86]Duhem, p. 90.

V

The earliest attempt by a member of the Academy of Sciences, after the arrival of Newton's *Opticks* and *Optice* in France, to verify Newton's conclusions by experiments, was evidently undertaken by the elderly Philippe de La Hire, at least according to the letter of Leibniz we quoted above. Leibniz, we saw, found the account which he had received unconvincing because of what he suspected to be La Hire's bias. That the elderly astronomer and mathematician had such *préventions* is not surprising, especially since he could scarcely have escaped from Mariotte's influence. He was close to and greatly admired the older man, even publishing posthumously his friend's *Traité du movement des eaux* in 1686.

There is no evidence that the Abbé Bignon was informed of, or followed, Leibniz's suggestion that some academician younger and with more leisure than La Hire and Malebranche should be assigned the task of confirming Newton's experiments. It was, in fact, a young provincial *savant*, not yet an academician, Dortous de Mairan, who—persuaded by his association with Malebranche—began to cite Newton's discoveries in his early publications and, if we may credit Pierre Coste, the translator into French of Newton's *Opticks,* was perhaps the first in France to carry out successfully some of Newton's key experiments on color.

Elsewhere I have treated in some detail this aspect of Dortous de Mairan's career, so a mere outline of the facts should suffice at this point.[87] Jean-Jacques Dortous de Mairan (1678–1771) was born in Béziers in the south of France to parents of the local gentry.[88] Educated at home until the death of his mother in 1694, at the age of nineteen he continued his studies at Toulouse. He then spent four years in Paris, where, through a relative of Malebranche, he came to know the famous philosopher and

[87]See my "Newtonianism of Dortous de Mairan," in Henry Guerlac, *Essays and Papers in the History of Modern Science* (Baltimore, 1977), pp. 479–490. For a summary account, see Chapter 3 pp. 65–66.

[88]For Dortous de Mairan see the *éloge* by the secretary of the Academy of Sciences, Grandjean de Fouchy; also J. Duboul, "Dortous de Mairan, étude sur sa vie et sur ses travaux," *Mémoires de l'Académie de Bordeaux* (1863, 2ᵉ trimestre), pp. 163–197.

his circle. Here he had his first introduction to the recent advances in mathematics and physical science, and his future scientific vocation was determined.

Mairan returned to Béziers in 1702 where he applied himself to his studies. We lose track of him until late in 1713, but in September of that year he embarked upon a philosophic correspondence with the venerable Father Malebranche in Paris.[89] In his first letter, dated from Béziers, 17 September 1713, Mairan explained that in the past year or two he had abandoned mathematics and physics in favor of the study of religious philosophy, guided by Malebranche's great book, and the writings of Pascal, Descartes, and Labadie. Lately he had been struggling with Spinoza, whose cogent reasoning deeply impressed him, but who led him inexorably to the most hazardous conclusions. Would the good Father be kind enough to point out the hidden fallacies in Spinoza's dialectic?

Malebranche replied, one senses a bit reluctantly, and the correspondence continued for nearly a year.[90] It was through this exchange that Mairan first learned about Newton's doctrine of light and color, for although Malebranche's letters dealt largely with metaphysics, he could not leave scientific questions aside. A letter from Malebranche, dated 12 June 1714, gave Mairan his first hint that the master had altered his earlier views concerning light. Here, in discussing the question whether, in the generation of plants and animals, a substance must be divisible to infinity, Malebranche remarked that he had treated this subject in an "optics which I gave in the last edition of the *Recherche de la vérité.*"[91] This led Mairan to reply that his curiosity was aroused to see the optical treatise Male-

---

[89]*Oeuvres de Malebranche*, XIX, 852–865, 870–879, and 882–912. The letters were published earlier in *Méditations métaphysiques et correspondance de N. Malebranche, Prêtre de l'Oratoire, avec J. J. Dortous de Mairan*, ed. Feuillet de Conches (Paris, 1841), pp. 93–177. Only the first four of the eight letters of the exchange appear in Joseph Moreau, ed., *Malebranche—Correspondance avec J.-J. Dortous de Mairan* (Paris, 1947).

[90]The letters, four by Dortous de Mairan with four replies by Malebranche, extend from Mairan's first letter of 17 September 1713 to Malebranche's last reply dated 6 September 1714, little more than a year before the sage's death (13 October 1715).

[91]*Oeuvres de Malebranche*, XIX, 886.

branche had mentioned. The edition of the *Recherche* he had used, Mairan added, is that of 1700 (that is, the fifth edition in which the Newtonian material does not appear); accordingly he has ordered the new edition of 1712 from Paris.[92] It would appear that soon after reading Malebranche's "XVIe éclaircissement" Mairan acquired a copy of Newton's *Optice*; for within a year, and often afterward, Mairan shows himself familiar, not only with Malebranche's modified theory of color, but also with the Latin version of Newton's *Opticks*. The first edition of Mairan's *Dissertation sur la glace,* published in Bordeaux in 1716, has several references to Newton's optical theories: to refraction, the transparency of bodies, and the explanation of color. One encounters this sentence: "For, according to the opinion of the illustrious M. Newton, the parts of nearly all bodies are naturally transparent, and their opacity comes only from the many reflections of these parts."[93] A note refers the reader to the Latin *Optice* of 1706, page 210. Further on appear the following remarks: "Now we know the extreme dependence of colors upon the force and the different rates of vibration of the subtle matter, or the rays of the sun. The excellent works that have appeared on the subject in recent years, no longer permit of any doubt."[94] The *Optice* is cited once again, as well as the discussion of light and color in the 1712 edition of Malebranche's *Recherche de la vérité.* Mairan's words, of course, echo the undulatory theory of Malebranche, rather than anything he could have learned from Newton's *Optice.* And finally we have this sentence: "This phenomenon is absolutely consistent with the general theory of refraction explained in the books I have cited above."[95] Here too a note refers the reader to Newton and Malebranche.

The fullest of Mairan's early references to Newton's doctrine concerning light and color is to be found in his essay on the cause of the light produced by phosphorescent substances. That study was published in Bordeaux in 1717 and, like the *Dissertation sur la*

[92]Ibid., p. 908, and *Méditations métaphysiques,* pp. 167–168.
[93]*Dissertation sur la glace* (Bordeaux, 1716), p. 78. This first edition was reprinted in Paris in 1717.
[94]Ibid., p. 79.
[95]Ibid., p. 83.

glace, it was awarded the gold medal of the Academy of Bordeaux. In it we find the following reasonably accurate account of the Newtonian theory of dispersion:

> Every kind of light has its definite refraction, that is to say, each color in passing from one medium to another, from air, for example, into crystal, is bent [*se rompt*] by an angle different from that of the other colors. That is what Mr. Newton, author of this discovery, calls the "different refrangibility of the colors of light." It was principally by this property that he found out all the others; and the ingenious experiments which he used to prove it, could serve by themselves to immortalize a name less celebrated than his.[96]

Then Mairan proceeds to recount in some detail Newton's classical experiments "in order to acquaint those who have not seen the *Opticks* of Mr. Newton, with what I shall have to say about this matter."

Unlike Malebranche, it would appear that Dortous de Mairan was converted to Newton's theory of light and color, not only by reading the *Optice* (and of course by the example and great reputation of his preceptor) but by assuring himself of its truth through repeating some of Newton's principal experiments. Our source for this assertion is Pierre Coste, a Huguenot refugee living in England who, while best known for translating into French Locke's *Essay concerning Human Understanding* (Amsterdam, 1700), will appear at the close of this narrative for his French version of Newton's *Opticks*. In his translator's Preface, after mentioning the successful repetition of Newton's experiments in Paris in 1719, to be treated shortly, Coste writes: "*M. de Mairan* les avoit aussi verifiées à Beziers en 1716" and apparently repeated them in 1717 "avec le même succez."[97] Even if Coste's testimony can be relied upon, and he could have got his information only at second or third hand, Mairan had no witnesses to his experiments and neither published

[96]*Dissertation sur la cause de la lumière des phosphores et des noctiluques* (Bordeaux, 1717), p. 48.

[97]*Traité d'optique* (Paris, 1722), Préface du traducteur. The reference to Mairan does not appear in the earlier Amsterdam edition (1720) of Coste's translation. See below pp. 144–147.

his results nor mentioned them in his writings. At this time he was little known, an obscure young *savant* of the provinces, and it is understandable that few heard of his experiments and that doubt persisted in influential quarters. It required heavier artillery than Mairan could provide to overcome the habitual and well-entrenched skepticism based on the failure of Mariotte, Philippe de La Hire, and doubtless others to confirm Newton's findings. The important figure, the man of great reputation who brought matters to a head, was none other than Leibniz.

Leibniz, we saw earlier,[98] mentioned in 1707 in a letter to Lelong the importance of settling the disagreement between Newton's results and Mariotte's. But he had raised the matter years before in letters to Huygens and to Fatio de Duillier.[99] It is clear that he was much concerned to have it decided once and for all, and at long last, whether Newton or Mariotte should be believed. Indeed, although the controversy over who invented the calculus, which began in 1705, had reached a peak of tension and ill temper (the *Commercium Epistolicum* which defended Newton was published in 1712), and although Leibniz had attacked Newton's theory of gravity in his *Théodicée* in 1710, we find Leibniz insisting once again that the question of the origin of prismatic colors should be settled conclusively.

In the *Acta eruditorum* for October 1713 appeared an anonymous review, almost certainly from the pen of Leibniz, of the third edition (London, 1710) of Samuel Clarke's Latin translation of

[98]See above p. 110 and n. 82.

[99]In a letter of October 1690 to Huygens, which seems not to have been sent, Leibniz wrote: "Je voudrois que vous eussiés voulû nous donner au moins vos conjectures sur les couleurs" (*Oeuvres de Huygens*, IX, 523). In January 1692, and again late in June 1694, Leibniz renewed his request (*Oeuvres de Huygens*, X, 229 and 651), but Huygens never complied. He died the following year. On 8 May 1694 Leibniz wrote from Hanover to Fatio in London: "Et quand à la lumière la grande difficulté à mon avis est de rendre raison des couleurs. . . . Je souhaiteray de sçavoir le sentiment de M. Fatio, et même celuy de M. Newton . . . sur les raisonnemens de M. Mariotte opposés en partie à M. Newton sur tout lorsqu'en accordant la différente refrangibilité, il nie pourtant, que la couleur ne paroist que par une separation des rayons colorés primitifs, croyant de pouvoir prouver que le même rayon change de couleur par la refraction" (Bibliothèque Publique et Universitaire de Genève. MS fr. 610. Papiers Fatio no. 10, fol. 24).

Rohault's *Traité de physique*.[100] In his notes to this Cartesian work, it will be recalled, Clarke (who not long before had translated Newton's *Opticks* into Latin) took pains to present in his notes the main tenets of Newton's natural philosophy; and at one point he gave a summary of Newton's discoveries and theories concerning light and color. Toward the end of this review these passages are singled out for special comment. Leibniz reminds the reader that objections that learned men "in France as well as England" raised against his theory were effectively refuted by Newton in various issues of the *Philosophical Transactions*. Many persons hope the opinion of this *vir perspicacissimus* might be freed from the objections effectively raised against him "by the most ingenious Mariotte, an indefatigable investigator of nature."[101] If it be true, Leibniz remarked, that a portion of the violet rays could be changed by a second prism into red and yellow, or red rays into blue and violet, as Mariotte claimed, then his experiments manifestly contradict Newton's. But Mariotte's experiments can be taken as convincing only if such a "modification" does in fact take place. In Leibniz's words: "To us, Mariotte's experiment will only appear decisive if indeed pure blue light changed into something else."[102]

This challenge, this virtual interpellation, could scarcely go unanswered. If Newton was aware that, in his first paper, his method of demonstrating the heterogeneity of white light was presented in an incomplete and ambiguous fashion, he seems to have been confident that in his *Opticks* the matter was set forth with sufficient clarity. Anxious to remove all doubts, he requested the gifted young experimenter, Jean Théophile Desaguliers (1683–1749) to "try the Experiment in the manner described" in Book I, Part I, Proposition IV of the *Opticks*, and presumably to publish the result.[103]

[100]*Acta eruditorum*, October 1713, pp. 444–448.

[101]Ibid., p. 447.

[102]"Nobis experimentum Mariotti tum demum videretur decisivum, si lumen coeruleum integrum in aliud mutatum fuisset" (ibid., p. 448).

[103]"Sir Isaac Newton . . . upon reading what has been cited out of the *Acta eruditorum*, desired Mr. Desaguliers to try the Experiment." (*Phil. Trans.*, 29 [1717], 435). Desaguliers in this article quotes some twenty lines from the review in the *Acta*. There is little doubt that its appearance was taken as a challenge by Newton.

This Proposition IV is really a short essay on the importance, if pure colors are to be obtained, of separating as far as possible the circular images he took to make up his prismatic image. These colored circles must be reduced in diameter, their centers remaining the same, if overlapping is to be minimized. Newton proposes various ways of effecting this: to reduce the size of the solar image by using an external screen outside the darkened room or, better still, by putting a biconvex lens in front of the prism and placing the targut screen at the focal distance of the lens. Experiment 11 (Book I, Part I, of the *Opticks*) with the accompanying figure, illustrates this procedure, but is not the experiment Newton asked Desaguliers to perform. It is worth noting that nowhere else in the *Opticks,* not even in the so-called *experimentum crucis* (Book I, Part I, Experiment 6) is a lens used for the purpose of reducing the overlapping of the solar image.

The experimenter Newton selected for this task was a happy choice. Born in La Rochelle in France in 1683, Desaguliers was the son of a Protestant pastor who two years later fled to the island of Guernsey to escape the consequences of the revocation of the Edict of Nantes.[104] Crossing to England, the father became minister of the French chapel in Swallow Street, London, and then moved to Islington where he kept a small school, assisted by his son. Desaguliers received his early education from his father and upon his father's death matriculated at Christ Church, Oxford, where he received the B.A. At Oxford he studied natural philosophy with John Keill: the first man, as Desaguliers later wrote, to teach Newtonian physics "by experiments in a mathematical manner." In 1710, when Keill departed for America, Desaguliers succeeded him as lecturer at Hart Hall, where he followed his master's

---

[104]For Desaguliers, see the sketch in the *Dictionary of National Biography* (by Robert Harrison) and in DSB (by A. R. Hall). Hall inadvertently has Desaguliers succeeding James Keill (the physician) rather than John, the older brother, as lecturer at Hart Hall in Oxford. There is a brief notice by E. G. R. Taylor in her *Mathematical Practioners of Tudor and Stuart England* (Cambridge, 1954), p. 300. I. Bernard Cohen discusses Desaguliers at some length (*Franklin and Newton,* passim) as an outstanding example of "experimental Newtonianism" and for his influence on Franklin. Cohen and Hall both cite Jean Torlais, *Un Rochelais grand-maître de la Franc-Maçonnerie et physicien au XVIIIᵉ siècle* (La Rochelle, 1937), which I have not seen.

methods, presenting the elements of hydrostatics, optics, and mechanics with well-chosen experiments.[105] It does not appear that Newton's experiments on light and color were among those shown to his auditors. Having taken deacon's orders and a wife, and receiving the M.A. in 1712, Desaguliers moved to London, settled in a house in Channel Row, Westminster, and in January 1713, began his course of demonstration lectures in natural philosophy "at Mr. Brown's, Bookseller, Temple Bar," in a sense as a younger rival of the already famous Francis Hauksbee.[106] Yet the courses the two men gave were markedly dissimilar; Hauksbee's experiments were more miscellaneous than Desaguliers's, less designed to illustrate fundamental principles, and more concerned to display his ingenuity and recent discoveries, many of them his own.[107] Nevertheless when Hauksbee died in April 1713, Desaguliers for a short time seems to have carried on Hauksbee's course to aid the widow. At the Royal Society, Hauksbee's death opened a new career to Desaguliers, whose experimental skill brought him to the attention of Newton and of Hans Sloane. In March of 1714 he was called upon to demonstrate his skill before the Fellows of the Society, and it is not without significance that in his first appearances he showed experiments on high-temperature thermometry like those which Newton had first described in 1701 in his anonymous *Scala graduum caloris*.[108]

It was apparent that Desaguliers could be counted on to repeat, with equal success, other experiments of Newton. He readily accepted the challenge of repeating Newton's optical experiments, and demonstrated a variation of the *experimentum crucis* and a num-

---

[105]These Oxford Lectures, based on Keill's, are embodied in Desaguliers's *Lectures of Experimental Philosophy* (2d ed., 1719) and their origin is described in the author's unpaginated preface. This was first published by Paul Dawson without the author's consent that same year as *A System of Experimental Philosophy*.

[106]The London lectures, in which the demonstration of Newton's optical experiments played a notable part, formed the basis of Desaguliers's more substantial *Course in Experimental Philosophy*, 2 vols. (London, 1730–1734).

[107]For Hauksbee see my "Francis Hauksbee—expérimenteur au profit de Newton," reprinted in Guerlac, *Essays and Papers*, pp. 107–119; also my sketch in DSB, VI (1972), 169–175.

[108]Journal Book, Royal Society of London, 18 March 1713/14 and 25 March 1714 (O.S.).

ber of Newton's other optical experiments "with Success before several Gentlemen of the Royal Society,"[109] not, it would seem, at a meeting of that body, but probably *chez* Brown, the bookseller, or in his own house. The Journal Book records the following item for 22 July 1714 (O.S.), with the president, Sir Isaac Newton, in the chair:

> An Account was read, drawn up by Mr. Desaguliers of the Experiments he made in order to the explaining and verifying of the Presidents Treatise of Colours which was ordered to be Entered and published in the Transactions.[110]

Desaguliers's paper, the first in a long succession of memoirs he was to publish on various aspects of experimental and applied physics, appeared in the *Philosophical Transactions* some three years after it was read.[111] His account of nine experiments, all of them repetitions of experiments described in Book I, Part I, of Newton's *Opticks,* is prefaced by an explanation of why this verification was deemed necessary:

> The manner of separating the primitive Colours of Light to such a Degree, that if any one of the separated Lights be taken apart, its Colour shall be found unchangeable, was not published before Sir Is. Newton's *Opticks* came abroad. For want of knowing how this was to be done, some Gentlemen of the English College at Liege, and Monsieur Mariotte in France, and some others took those for primitive Colours, which are made by immitting a Beam of the Sun's Light into a dark Room through a small round Hole, and refracting the Beam by a triangular Prism of Glass placed at the Hole. And by trying the Experiment in this manner, they found that the Colours thus made were capable of change, and thereupon reported that the Experiment did not succeed. And lately the Editor of the *Acta Eruditorum* . . . desired that Sir *Is. Newton* would remove this Difficulty.[112]

[109]*Phil. Trans.*, 29 (1717), No. 348, 435.
[110]Journal Book, Royal Society of London, for 22 July 1714 (O.S.).
[111]"An Account of some Experiments of Light and Colours, formerly made by Sir Isaac Newton, and mention'd in his Opticks, lately repeated before the Royal Society by J. T. Desaguliers, F.R.S., "*Phil. Trans.*, 29 (1717), No. 348, 433–447.
[112]Ibid., p. 433. No such explicit request was actually made in the review, but the inference was obvious. See n. 102.

As described in the *Transactions*, the experiments Desaguliers reported to the Royal Society differed in significant ways from Newton's descriptions in the *Opticks*. Desaguliers's paper is accompanied by a folding plate illustrating nine experiments with twenty-four schematic figures. Especially notable is the care with which the apparatus and its disposition are described. It would appear that Newton's casualness in this respect, the cause of the misunderstandings and controversies that demonstrated the need for further elaboration of detail, had driven home the lesson. Desaguliers's account of these experiments is an early example of that painstaking description of the conditions and disposition of an experiment now expected in reports of scientific results so that they can be readily confirmed or "falsified."

The first experiment Desaguliers described is essentially a repetition of the early one in which Newton examined through a prism rectangles of paper painted half red and half blue. In one experiment Desaguliers used, instead of the colored paper, two pieces of ribbon sewn together, and in another, "skeens" of red and blue worsted.[113] He was at pains to vary the kinds of prisms he used, made of different sorts of glass and with different refracting angles. The spectra, he noted, were sharper and the colors more distinct when he used a prism of greenish glass "such as Object Glasses of Telescopes are made of," for these were free from the veins encountered in the ordinary prisms of white glass "by which the Colours are too much thrown into one another."[114] There are difficulties even in the experiments using ribbon or worsted. The effect is not produced clearly if the background is even slightly illuminated; it should consist of a black cloth placed so "that no Light falling upon it can be reflected so as to disturb the Phenomenon."[115] Desaguliers also repeated Newton's experiment

[113]Compare *Opticks*, Book I, Part I, Proposition 1, Theorem 1. In perhaps his earliest recorded experiment on color, described by A. R. Hall in "Sir Isaac Newton's Note-Book 1661–65," *The Cambridge Historical Journal*, 9 (1948), 247–248, Newton used a blue and a red thread sewn together, just as Desaguliers used colored ribbons.
[114]Desaguliers, "An Account of some Experiments," pp. 436 and 437.
[115]Ibid., p. 437.

using two holes in the window shutter and two prisms so as to produce two spectra, one above the other on a sheet of white paper, so that the red end of one spectrum just touched the purple, as he calls it, of the second. Looking at the two contiguous spectra with a third prism nine feet away, Desaguliers found the red portion distinctly displaced from the purple.[116]

A repetition of Newton's apparently simple experiment of producing an elongated spectrum by means of a single prism is described in great detail.[117] Because his perfunctory description of this experiment in the famous first paper of 1672 had caused great confusion among his Continental readers, Newton was careful to describe it with considerable detail in his *Opticks*. Here he gave the exact size of the hole in the window shutter, the precise orientation of the prism, its distance from the white paper screen on the wall, the diffracting angles of the prisms he used, and so on.[118] Desaguliers can hardly be said to have improved upon Newton's exposition in the *Opticks*. He adopts most of the precautions that Newton enumerated, making only one modest addition: where Newton merely said that he placed the glass prism "at a round Hole" in the window shutter, Desaguliers specified that his prism was placed five inches from the hole.[119]

In his Experiment VI Desaguliers then recounts his success with the striking experiment, described, and indeed illustrated with a figure, in Newton's first paper.[120] Here a convergent lens was used to bring the separated rays of the spectrum to a focus, showing that the colors became less distinct as one inserted a screen progressively nearer the focus, at which point the image appears white, and

---

[116]Experiment II, ibid., pp. 437–438. See the *Opticks,* Book I, Part I, Proposition I, Experiment 7. Newton commonly called the color at the high frequency end of the visible solar spectrum *violet* or (as in his paper of 1672) *violet-purple.* He eventually decided to call *purple* that color which is not found in the prismatic spectrum, but can be produced by mixing red with violet or blue lights. See *Opticks,* Book I, Part II, Proposition IV. But Desaguliers used purple to mean the portion of the blue verging on the violet, what Newton called indigo.

[117]Desaguliers, Experiment IV, pp. 439–442.

[118]*Opticks,* Book I, Part I, Proposition II, Experiment 3.

[119]Desaguliers, Experiment IV, p. 440.

[120]In the *Opticks* this experiment (Book I, Part II, Experiment 2) is described in as much or more detail.

beyond which the rays diverge again to form a spectrum with the order of the colors reversed. Desaguliers, however, tried the experiment with lenses of different radii, and warned that "Care must be taken that the very end of the Red, and the Extremity of the Violet be taken in by the *Lens*; otherwise the *Spectrum* will not be perfectly white."[121]

At this point, Desaguliers turned his attention to Newton's *experimentum crucis,* which he carried out with some significant modifications.[122] Light from a hole in the window shutter passes through a first prism and is intercepted by a board placed on the ground and propped up at an angle. Immediately behind the board, which is pierced by a hole a quarter inch in diameter, he placed a second prism. For convenience in rotating the prism, he fixed both ends "into a triangular Collar of Tin, from the end of which came a Wire which was the Axis of the Prism produc'd; and so laid it on two wooden Pillars, with a Notch on the Top to receive the Wires, and fix'd it to a small Board just broad enough to stand fast."[123] The rays from the second prism were projected on the ceiling, and by rotating the first prism he could observe the varying extent to which the several rays were refracted by that prism. Desaguliers emphasizes that the orientation of the second prism is critical: "Several have confess'd to me that they at first us'd to fail in this Experiment, for want of setting the second Prism in a due Inclination."[124]

But there are more subtle sources of error, as this very experiment showed. Whereas the colors projected on the ceiling

[121]Desaguliers, Experiment VI, pp. 442–443. Like Newton, the first to use the word in English, Desaguliers meant by "spectrum" simply the solar image, whether round and white or extended and colored. For neither man did it imply a gradation or scale of colors but only an insubstantial image. See my note, "The Word *Spectrum*: A Lexicographic Note with a Query," *Isis,* 56 (1965), 206–207. For a possible classical use of the word, see my "Augustan Monument: The Opticks of Isaac Newton," in Guerlac, *Essays and Papers,* p. 165, n. 24.

[122]First described succinctly in Newton's classic first paper on light and color, *Phil. Trans.,* 6, no. 80, February 19, 1671/72, pp. 3078–3079. See above pp. 92–93. It is more clearly set forth in the *Opticks,* Book I, Part I, Proposition II, Experiment 6, where Newton does not use the phrase *experimentum crucis* to describe it.

[123]Desaguliers, Experiment VII, pp. 443–444.

[124]Ibid., p. 444.

appeared to the naked eye to be pure and unchanged by the second refraction, when Desaguliers looked at them through a third prism "they afforded new Colours," owing to their not being fully separated. Only when the prisms are of good quality, such as he and Newton recommended, and when there are no clouds near the sun, will you get unmixed colors in the extreme regions of the red or the violet: "otherwise not."

Even so, as Desaguliers readily acknowledged, indeed stressed, in the introduction to his paper, it is a difficult matter to obtain rays of pure, unmixed color. Mariotte, he remarked, had observed red and yellow colors in the supposedly pure violet ray passed through his second prism, not because the ray had been "modified" by that prism, but because the spectral colors had not been sufficiently separated, or because of scattered light coming from bright clouds near the sun. Echoing a suggestion of Newton's, he described how this could be corrected by placing a screen in the open air some ten or twenty feet from the window.[125] Pierced by a small hole, round or oblong, "not above one eighth or one tenth part of an Inch broad," the screen will intercept "not only the bright Light of the Clouds next encompassing the Sun's Body, but also the greatest part of the Sun's Light. For thereby the Colours will become less mixed." With this arrangment the beam then is passed through the hole into the darkened room, and through a 60° prism placed parallel to the oblong hole. "In this manner," Desaguliers adds, "the Experiment may be tried with Success, but the Trial will be less troublesome if it be made in such manner as is described in the fourth proposition of the first Book of Sir Is. Newton's *Opticks*."[126]

Now it is to be noted that the experiment to which Desaguliers referred (*Opticks*, Book I, Part I, Proposition IV) describes an arrangement which Newton did not use in the *experimentum crucis* as reported either in the first paper of 1672 or in the *Opticks* (Book I, Part I, Proposition II, Experiment 6). Nor in this so-called crucial

[125]Ibid., p. 434. Newton's suggestion of an external screen is found in the discussion of Proposition IV, Problem I, in Book I, Part I, of the *Opticks* (2d and 3d eds., pp. 56–57).
[126]Desaguliers, p. 435.

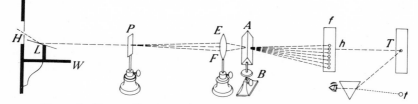

Figure 1. Desaguliers's version of Newton's *experimentum crucis.*

experiment did Newton mention using the device of the outdoor screen. Indeed, in repeating this experiment Desaguliers followed Newton's description closely with only trivial modifications. Yet, as he testified, this particular arrangement did not yield truly monochromatic rays "for which reason," he added, "I made the following Experiment, to prove that if the Colours be well separated, they are truly homogeneal and unchangeable."[127]

This experiment, so performed as to obviate the difficulties that had led to the results reported by Mariotte, has an arrangement worked out by Desaguliers, although largely composed of elements suggested by Newton in his Proposition IV; it is, in fact, the *experimentum crucis* modified by the use of an indoor screen at the window and of the lens recommended by Newton.[128] Since this is probably the first time that this experiment was performed so as to produce nearly "pure" rays, it deserves a careful description.

Desaguliers made a hole (H) in his window shutter 2 inches wide and covered this hole with a moveable tin plate which let in a tiny beam of light through its own hole 1/16 inch in diameter (see Figure 1). On a board or bracket (W) on the inside window sill he placed a plane mirror (L) inclined so as to reflect the beam of light to the other extremity of the room. To reduce any scattered light from the irregularity of the mirror, he interposed a frame of pasteboard (P) pierced by a 1/16 inch hole. A biconvex lens (EF) with a radius of 4 1/2 feet, held on an adjustable stand, intercepted the

[127]Ibid., p. 444.
[128]Proposition IV, Problem 1, in Book I, Part I, of the *Opticks* (2d and 3d eds., pp. 57–59).

125

beam at a distance of 9 feet from the mirror so that the image of the hole in the pasteboard appeared on a screen of white paper (f) 9 feet from the lens. Beyond the lens, and as close to it as possible, he placed a prism (A) upright on its adjustable stand (B), and so supported that it could be rotated about its axis. This produced on the screen an elongated spectrum 30 or 40 times its breadth.

> The Colours in this Case were very vivid and well separated, only the Violet had some pale Light darting from its End, upon account of some Veins in the Prism A, and the Light not coming directly from the Sun, but reflected; which ought not to have been, if the Sun had been low enough to have thrown the Rays a good way into the Room without the Help of a Looking Glass.

To show that the spectral colors were "simple and homogeneal Lights," Desaguliers then made a hole (h) in the paper which received the colored spectrum, and allowed the red rays to pass through, and impinge on a second paper screen (T). The image on this second screen appeared "red, round and unchang'd" both to the naked eye and when looked at through prisms of different refracting angles. And Desaguliers concluded: "I made the Experiment upon all the Colours, which by this means appear'd simple and homogeneal." Here again he cautioned that without proper optical equipment the experiment would not succeed:

> The Lens ought to be very good, without Veins or Blebs, and ground to no less a Radius than I mentiond in the Experiment; tho' a Radius of a Foot or two longer is not amiss. The Prism ought to be of the same Glass as the Object-Glasses of Telescopes,[129] the white Glass, of which Prisms are usually made, being commonly full of Veins. And the Room in these last Experiments ought to be very dark.

Desaguliers evidently was reporting these experiments as he had first performed them in private, for in the next-to-last paragraph of the published account we read:

[129]That is, the greenish glass mentioned above, p. 121.

A few Days after, having got very good Prisms made for the purpose of the above mention'd Glass, I made all the Experiments over again before several Members of the *Royal Society* with better Success; and had the *Spectrum* very regularly terminated, without any pale Light darting from the Ends of it.[130]

We may well pause at this point to ask some presumptuous questions: Did Newton, in fact, having conceived the much-discussed *experimentum crucis*, and having satisfied himself by its means that white light can be decomposed into rays "differently refrangible," actually perform the experiment in such a way as to prove the intimate relation of refrangibility to color? Did he, in other words, ever obtain unmixed, "monochromatic" rays? Could he have done so? It would appear that if he followed his own description of the *experimentum crucis* (*Opticks*, Book I, Part I, Experiment 6), the result would have been as inconclusive in this respect as Desaguliers, who followed Newton's instructions closely, found it to be. Montucla was right when he suggested that in the *Opticks* Newton for the first time "disclosed the manner of decomposing light sufficiently to produce unalterable colors."[131] Yet, taken by itself, the description of the *experimentum crucis* would not have sufficed; it was necessary, as Desaguliers's experiment makes clear, to improve the original disposition with refinements like those suggested in the *Opticks*, Book I, Part I, Proposition IV. That the description of the *experimentum crucis* itself in Newton's book was inadequate is tacitly admitted by Desaguliers in the concluding paragraph of his paper, where he says that for further experiments one should consult the *Opticks*:

> to which I might have referr'd the Reader altogether; but that I was willing to be particular in mentioning such things as ought to be avoided in making the Experiments above-mention'd; some Gentlemen abroad having complained that they had not found the Experiments answer, for want of sufficient Directions in *Sir Isaac Newton's Opticks*; tho' I had no other Directions than what I found there.

[130]Desaguliers, p. 447.
[131]Montucla, II, 623.

In the book, yes; but not in the description of the *experimentum crucis*. We should not, I think, deal too harshly with Mariotte who had only Newton's first papers to guide him, or with those later "Gentlemen abroad," whoever they were, who did not find "sufficient directions" even in the *Opticks* for performing the experiment in a manner sufficient to prove the inalterability of the separated rays, and the one-to-one correspondence of color to refrangibility.[132]

VI

Desaguliers's experimental confirmation of Newton's results played a central role in convincing the doubters on the Continent. Political and, as it happened, astronomical events conspired to spread abroad the news of his success. The year 1715—with which, to all intents and purposes, the eighteenth century was ushered in—was an eventful one for England as it was for France. Two years had elapsed since peace was signed at Utrecht (April 1713); and the death of two rulers, Queen Anne in August of 1714, and the aged enemy of England, Louis XIV, the following year, marked the end of an era. In September of 1714 the first Hanoverian monarch landed in England.

The accession of George I brought embassies of congratulation from the chief countries of Europe, and in their train, disguised as secretaries of mission, came a small platoon of scientific personages curious to see the great Newton, to attend meetings of the Royal Society, and—for at least one of them this was the principal attraction—to observe an eclipse of the sun which promised to be visible in its totality at London, on 22 April 1715 (O.S.), the first, according to Edmond Halley, to be visible in London since the year 1140.

As secretary to the Dutch embassy came W. J. 'sGravesande, aged twenty-seven, and destined to be Newton's earliest Dutch exposi-

---

[132]Desaguliers, p. 447. Desaguliers supplemented these Newtonian experiments by one of his own devising, performed before the Royal Society on 13 January 1714/15 and printed in the *Phil. Trans.*, 29 (1717), No. 348, 448–452 with the title "A plain and easy Experiment to confirm Sir Isaac Newton's Doctrine of the different Refrangibility of the Rays of Light."

128

tor. Arrived on English soil, 'sGravesande met the sons of Gilbert Burnet, who had been his fellow students at the University of Leiden and who introduced him into London scientific circles.[133] He came to know Newton and became especially close to Desaguliers, whose demonstrations he watched, and with whom he collaborated on experiments. He may well have been present when, for the Fellows and guests, Desaguliers repeated (in February 1715) the second experiment of Newton's *Opticks*;[134] later that month 'sGravesande's friend William Burnet proposed him as a Fellow of the Society.[135] That 'sGravesande acquired, during his year in England, notably from Desaguliers, full mastery of Newton's optical experiments cannot be doubted: they are described with care in his famous work of experimental Newtonian physics, published in Latin in 1720–1721.[136] This work was translated by Desaguliers, who singled out for approval 'sGravesande's demonstration of Newton's doctrine of light and color, "proved by the most considerable of his Experiments, which Dr. 'sGravesande performs with an Apparatus very ingenious contrived, and nicely expressed [i.e., illustrated] by curious Figures."[137]

[133]The envoys of the States General were Borsele van den Hoge and Baron Wassenaer van Duyvenvoorde. The latter, impressed by the reception 'sGravesande received from Newton and other British scientists, was instrumental in obtaining for his young colleague in 1717 the post of Professor of Mathematics and Astronomy at the University of Leiden. See J. N. S. Allamand's "Histoire de la vie et des ouvrages de M^r 'sGravesande," prefaced to his *Oeuvres philosophiques et mathématiques de 'sGravesande* (Amsterdam, 1774). This is the principal biographical source; but see also Pierre Brunet, *Les physiciens hollandais et la méthode expérimentale en France au XVIIIe siècle* (Paris, 1926), chap. 1, and the sketch by A. R. Hall in DSB, V (1972), 509–511.

[134]"Mr. Desaguliers shewed ye 2d Experiment of the first part of Sir Isaac Newton's Opticks which succeeded to the satisfaction of all those that were present" (Journal Book, 3 February 1714/15).

[135]Journal Book, 24 February 1714/15. He was elected on 9 June 1715. See *Record of the Royal Society of London*, 4th ed. (London, 1940), p. 393. How long the mission stayed is not clear, but 'sGravesande's visit lasted more than a year, for on 2 February 1715/16 he announced to the Society his imminent departure, asking if he could be of service while in Holland. See the Journal Book for that date.

[136]*Physices elementa mathematica experimentis confirmata. Sive introductio ad philosophiam Newtonianam*, 2 vols. (Leiden, 1720–1721).

[137]*Mathematical Elements of Natural Philosophy, Confirm'd by Experiments; or, An Introduction to Sir Isaac Newton's Philosophy. Written in Latin by William-James 'sGravesande, Translated into English by J. T. Desaguliers*, 4th ed., 2 vols. (London, 1731), II, v–vi.

The delegation of French scientists, at least one of whom came especially to observe the solar eclipse, included the botanist and chemist C. J. Geoffroy (1685–1752), the younger brother of the man to whom Hans Sloane had several years before sent a copy of the *Opticks*; the mathematician, Pierre Rémond de Montmort; and an accomplished astronomer, the Chevalier de Louville. All three were members of the Royal Academy of Sciences in Paris, and all had been introduced well before their departure by letters of recommendation from the elder Geoffroy to Hans Sloane.[138] They were accompanied by the Paduan aristocrat, Antonio de Conti, whose attempts to mediate the Newton-Leibniz dispute over the invention of the calculus were the main fruits of his visit and his claim to a shred of immortality. All four men were cordially received by the Royal Society as honored guests, and were admitted as Fellows in June of 1715.[139]

In the letter he wrote to Hans Sloane from Paris on 17 April 1713 recommending his brother and M. de Montmort, Geoffroy hoped that both men could be admitted to a meeting of the Royal Society, and asked that Sloane arrange for them to see the experiments of Francis Hauksbee with his air pump and "those concerning colours reported in Mr. Newton's book of optics if it is possible to see them."[140] Desaguliers records that he did in fact demonstrate an experiment on light and color for the benefit of "Monsieur Monmort [*sic*] and others of the Royal Academy of Sciences," presumably the Chevalier de Louville and the younger Geoffroy.[141]

[138]See British Library, Sloane MSS 4044, fols. 30–31 and 35.

[139]Louville, Montmort, and the younger Geoffroy, together with 'sGravesande and a number of others, were elected on 9 June 1715 (*Record of the Royal Society*, p. 393). The three Frenchmen had been proposed to the Society on 5 May by Newton himself (Journal Book for that date). Conti was elected in November.

[140]Sloane MSS 4044, fol. 35r and v. Hauksbee, it will be recalled, died the very month this letter was written.

[141]Desaguliers, p. 435. Coste wrote later that Desaguliers, following the method "décrite au long" in Proposition IV of Book I of the *Opticks*, "fit voir distinctement en 1715 à Londres à M. *Remond de Montmor* [*sic*], M. le Chevalier de *Louville*, & autres Membres de l'Academie Royale des sciences," that well-separated rays are absolutely inalterable as to their color and refrangibility. See his "Préface du traducteur" to Newton's *Traité d'Optique* (Paris, 1722). At the time of his visit to England Montmort was not yet, as our sources (Desaguliers, Halley, and Pierre Coste) all suggest, a member of the Academy of Sciences. He became *associé libre* early in 1716. Louville, on the other hand, had been made *associé astronome* in March 1714.

Apart from the repetition of one of Newton's experiments early in February 1714/15, mentioned above, I have not found when other experiments were performed for the foreign visitors, but it was probably at some meeting of the Royal Society before 22 April 1715 (O.S.), the day of the solar eclipse.

Preparations for observing the eclipse were entrusted to the sure hands of the distinguished astronomer, Edmond Halley. On orders of the Society, as he put it, he set up at Crane-Court in Fleet Street, a meeting place the Society had acquired in 1710, a thirty-inch quadrant with telescopic sights, several other telescopes for additional observers, and a "very good" pendulum clock adjusted to mean solar time. When the day came, conditions for observation were perfect, the sky a serene azure blue. Some men grouped themselves around the observing instruments in the courtyard, while others took up positions on the roof, where they could see the horizon and observe the stars and planets suddenly visible during totality.[142] A great many Fellows and visitors gathered on this day, although Halley, in his report, singles out for special mention Lord Parker, not yet a Fellow but an enthusiastic amateur of astronomy and optics, who took an active part in the observations.[143] And Halley notes that on hand to observe this great event, to observe the striking corona and solar prominences he later described so well, "were also present several foreign gentlemen, and among them *le Chevalier de Louville* and M. Monmort, both members of the *Royal Academy* of *Sciences* at *Paris*: the first whereof came purposely to view this Eclipse with us," and carried out his share of the observations.[144]

Louville's account is more anecdotal than Halley's. He tells us, for example, as indeed Halley does, that immediately after the

---

[142]"Observations of the late Total Eclipse of the Sun on the 22d of April [O.S.] last past, made before the Royal Society at their House in Crane-Court in Fleet-Street, London. By Dr. Edmund [*sic*] Halley, Reg. Soc. Secr. With an Account of what has been communicated from abroad concerning the same," *Phil. Trans.*, 29 (1717), No. 343, 245–262.

[143]George Lord Parker, later Earl of Macclesfield, became F.R.S. in 1722 and was president of the Royal Society from 1752 to 1764, cosseting the bill on the reformed calendar through the Peers in 1751. For his astronomical work, and his association with James Bradley, see the DSB, II (1970), p. 388.

[144]Halley, p. 251.

eclipse, the sky clouded over, which led the Frenchman to comment on the climate of Britain and to remark that in the month he had spent in England he encountered only three clear days. He describes how, during totality, owls flew by, obviously believing that night had fallen.[145]

Of Louville and Montmort something more should be said, for both were linked closely, although in different ways, with the subsequent adoption of Newton's views in France. It is doubtless because of what they observed on their English visit, and at their urging, that Newton's experiments on light and color were successfully repeated in France for the first time.

Jacques-Eugène d'Allonville, Chevalier de Louville (1671–1732), as astronomer of some repute,[146] should be more often remembered than he is, for he was the first member of the Royal Academy of Sciences to publish an astro-physical memoir based on Newtonian principles. Robert Grant cites him, not for this, but for his pioneering work in applying the micrometer to the eyepiece of a telescopic quadrant, a device that served him well in his specialty of determining with high precision the diameter of the sun.[147] Described in the *Mémoires* of the Academy of Sciences for 1714, the device was employed by Louville during the London solar eclipse,

[145]For Louville's account, "Observations faites à Londres de l'éclipse totale du soleil du 3 May 1715, nouveau stile," see the *Mém. Acad. Sci.* for 1715 (1717), pp. 89–99.

[146]The main biographical source for Louville is the *éloge* by Fontenelle in *Oeuvres de Fontenelle,* new ed., 8 vols. (Paris, 1790–1792), VII, 435–444, from which the sketch in Didot-Hoefer, *Nouvelle biographie générale* is largely derived. But see also A. R. Hall, "Newton in France," pp. 239–241.

It is tempting to identify the Chevalier de Louville with that staunch Newtonian, the mysterious "M. le Chevalier," whose opinions, in the letters of 1718, were reported by Father Laval. The chronology is suggestive, for the letters would have been written not long after Louville's return to Paris in 1715. Suggestive, too, is his astronomical visit in 1713 to Marseilles where Laval had his observatory, as well as the familiarity of "M. le Chevalier" with Newton and Halley and with the optical discoveries Louville saw demonstrated in London. There are arguments against this identification, such as Laval's description of him as a "capitaine de vaisseaux" having a brother who was a bishop. Louville was a younger son, himself destined at first for the clergy, and the only brother we know of is the diplomat Charles-Auguste, marquis de Louville, some three years his elder.

[147]Robert Grant, *History of Physical Astronomy* (London, n.d. [1852]), p. 481. A. Wolf, *History of Science, Technology, and Philosophy in the Eighteenth Century* (New York, 1939), mentions it (pp. 126–127) and also refers to Louville's invention of a portable transit instrument (p. 132).

so Halley tells us, to observe the occultation of three spots then visible on the sun.[148] Even before his 'teens, Louville was attracted to mathematics; his formal education completed, he entered the navy and took part in the disastrous battle of La Hogue in 1690, then transferred to the army—the navy had been virtually decimated—and was captured at Oudenarde. With peace established by the Treaty of Utrecht, he left the service and devoted himself to astronomy. In 1713 he traveled to Marseilles to make a careful measurement of the height of the celestial pole, so as to correlate his observations with those of Pythias of Massilia and to ascertain whether, as he believed, the obliquity of the ecliptic had decreased since Antiquity. The following year he entered the Academy of Sciences as *associé astronome*, becoming a full member (*pensionnaire*) in 1719. The next year he presented to the Academy his memoir on the *Construction et théorie des tables du soleil*, basing the exposition and calculations on Newton's theory of gravitation.[149]

Louville, unless a crucial document is misdated, seems to have visited England at least once before his trip to observe the eclipse; Fontenelle, secretary of the Academy of Sciences, wrote to Newton on 29 May 1714 thanking him on behalf of the Academy for "un Recueil de différentes piéces de vous, qu'elle a reçu des mains de M. le Chevalier de Louville."[150] This is quite possible, for the elder Geoffroy's letter of introduction to Hans Sloane is dated, as we saw, 17 April 1713. When Louville wrote in 1715 of having spent a month in England, during which he saw only three clear days, this may refer to this earlier visit, for the visit of 1715 was perhaps longer than that. From the Society's Journal Book it is evident that Louville had returned to France by 14 July 1715, for on that date

---

[148]Halley, pp. 251–252.

[149]"Construction et théorie des tables du soleil," *Mém. Acad. sci.* for 1720 (1722), pp. 35–84. See Arthur Berry, *Short History of Astronomy* (Dover reprint, 1961), p. 290. Brunet, though admitting that the idea of gravitational attraction supplied the basis for Louville's calculations describes the idea as "timidement présentée." See his *Introduction des théories de Newton*, p. 85. Hall, "Newton in France," pp. 239–241, mentions that in Louville's gravitational work Newton is not mentioned, and Newton's explanation of planetary motion is ascribed to Kepler!

[150]Fontenelle to Newton, 20 May 1714, in Newton *Correspondence*, VI (1976), 145–146, no. 1084, previously published by G. Bonno, "Deux lettres inédites de Fontenelle à Newton," *Modern Language Notes*, 54 (1939), 188–190.

the Society received a Latin letter expressing thanks for his election to that body and proposing to discover whether the moon has an atmosphere by observing the change of color of stars as they disappear behind the moon or reappear after occultation.

Pierre Rémond de Montmort (1678–1719) is an equally interesting personage, better known than Louville to students of Newton's science.[151] Born into that privileged caste of the *noblesse de robe*, and destined for a career in the law, he fled France and paid his first visit to England, so Fontenelle tells us, to escape the drudgery of a legal education. In the course of his travels, which took him to Holland and Germany, a copy of Malebranche's *Recherche de la vérité* fell into his hands and determined then and there his vocation of philosopher and mathematician. Returning to Paris in 1699, he became a close friend and disciple of Malebranche and an intimate of the philosopher's circle. He learned the rudiments of geometry and algebra from Carré and Guisnée; and in the company of a gifted young friend, a man named Nicole, he plunged into the mysteries of the calculus, then being unraveled by the Bernoullis, as well as by the Marquis de l'Hospital, Varignon, and others of Malebranche's circle. In 1700 Montmort paid a second visit to England, and on this occasion made Newton's acquaintance and visited Newton's Oxford disciple, David Gregory. After returning to France he bought the fief of Montmort near Epernay in Champagne, where he lived henceforth and carried out his mathematical investigations. In 1706, or at the latest 1707, Montmort brought out a privately printed separate of Newton's *De quadratura curvarum,* the work on integration which was one of the two Latin appendices published in the *Opticks* of 1704 and the *Optice* of 1706. Could it have been this separate, rather than Newton's book, that Malebranche studied in the late summer and autumn of 1707, for Montmort seems to have been Malebranche's host during this visit to the country?[152]

[151]For Montmort see Fontenelle's *Oeuvres* (1790–1792), VII, 45–60; also Ian Hacking's sketch in DSB, IX (1974).

[152]See pp. 109–110 and n. 81.

Montmort did not lose contact with Newton in later years. In 1708 he published his pioneer treatise on probability of games of chance, the *Essay d'analyse sur les jeux du hazard,* and promptly dispatched a copy to Newton through the good offices of the Abbé Bignon, president of the Academy of Sciences. In the accompanying letter, dated 16 February 1709, Montmort confessed that two years earlier he had caused to be printed in Paris some hundred copies of the *De quadratura curvarum* for distribution to the *savants* of France.[153] Again, during the visit of 1715, Montmort saw something of Newton, was received in his household and became an ardent admirer of the beauty and accomplishments of Newton's niece, Catherine Barton. On this visit he came to know Abraham De Moivre, whose *Doctrine of Chances* owed something to the *Essay d'analyse,* and above all Brook Taylor, at that time a secretary of the Royal Society, with whom Montmort subsequently corresponded, and whom he received in Paris. After his return to France in 1716, Montmort sent a hamper of champagne to Taylor, with the request that fifty bottles be sent to Newton and Mistress Barton. Perhaps he doubted that it would reach its destination, for he wrote: "Ce seroit dommage que ce bon vin fut bu par des commis de vos douanes: étant destiné pour des bouches philosophiques, et la belle bouche de Mademoiselle Barton."[154]

Montmort is best known for the sensible and sensitive position he took, as expressed in letters to Brook Taylor, concerning the dispute over the invention of the calculus. Unlike Louville, and despite the admiration he frequently expressed for Newton and the favor he enjoyed of being admitted into his home, Montmort never, if we are to believe Fontenelle, accepted the controversial notion of universal attraction.

[153]Newton *Correspondence,* IV (1967), 533–534, no. 752. Columbia University's copy is bound at the end of Montmort's *Essay d'analyse.* Montmort's printer could have followed either the English *Opticks* of 1704 or the Latin *Optice* of 1706; it is difficult to tell. But the pagination, signatures, and catchwords in Montmort's version differ from both. Its title page (not strictly speaking a half title) simply copies the title page introducing the *De quadratura* in both editions. It bears no date or imprint, and Newton's name nowhere appears.

[154]Sir David Brewster, *Memoirs of the Life, Writings, and Discoveries of Sir Isaac Newton,* 2 vols. (London, 1855), II, 491. See also the same, p. 436.

But we can hardly doubt that, like Louville, he was persuaded of the correctness of Newton's views concerning light and color. Already predisposed by his association with Malebranche to accept the findings in the *Opticks*, he must have returned to France utterly persuaded by the successful demonstrations of Desaguliers. And we can conclude, without serious risk of error, that it was the informal reports of these three academicians, warning of the precautions that must be observed, which led, under more or less official auspices, to successful repetitions of these famous experiments in France.

Even before these experiments were carried out, we have a trace here and there of a changed attitude in France toward Newton's work on color. A trivial index of this new shift of opinion is found by comparing the first two editions of the *Expériences de physique* of Pierre Polinière, a pioneer lecturer in experimental physics.[155] The book of this mediocre precursor of the Abbé Nollet appeared in a single-volume duodecimo in 1709 and proved extremely popular. An expanded second edition was published in 1718; the fifth and last, in two volumes, is dated 1741. The changes that Polinière made in the sections dealing with light and color, between the first and second editions, are of special interest, for they record the penetration of Newton's optical discoveries into the wider world of French popular science.

In the first edition of 1709, after describing the production of spectral colors with an equiangular prism, Polinière explains:

> One should remember [*Il faut considerer*] that the different colors consist only in certain modifications of light which make as many differ-

[155]Polinière, born in 1671 in Normandy, completed his humanities at Caen. Coming to Paris for the course in philosophy, he studied mathematics with Varignon, then earned a medical degree. He published in 1704 an ill-received *Elémens de mathématiques*; but his real career began with the primitive course in experimental physics which he undertook shortly after, and which consisted in demonstration lectures given to students toward the end of their year of philosophy in the various colleges of the University of Paris and in Jesuit colleges. For Polinière see the sketch by David Corson in DSB, XI (1975), 67–68; Didot-Hoefer, *Nouvelle biographie générale*, and the short prefatory sketch in the second volume of his *Expériences de physique*, 5th ed. (1741). He is mentioned briefly in Brunet's *Physiciens hollandais*, pp. 101–102.

ent impressions on our eyes. The motion of light can be changed and determined in a great number of different ways, according to the diversity of the surfaces of reflecting bodies, and according to the different ways it passes through transparent bodies.[156]

As this quotation indicates, Polinière, in this first edition, totally ignores Newton and his discoveries; his vague account merely echoes the older, and still widely accepted, "modification" theory of the production of color.

But this section of his book was markedly altered when Polinière prepared his second edition, "revue & beaucoup augmentée," of 1718. After describing the method of producing prismatic colors he writes:

One of the English scientists has thought so long about this experiment, and has elaborated on it so fully, with the addition of a great number of other experiments that he claims to have demonstrated that light is composed of a multitude of rays with different properties. Among these rays are those that produce the sensation of red, others the sensation of yellow, still others the sensation of other colors.[157]

In a marginal note, Polinière gives his source as Newton's Latin *Optice* of 1706.

Although Polinière often seems confused and to have lingering doubts, by 1718 he had abandoned, and without hesitation, his modification theory of color. Where in the first edition he had written the phrase "light can be still further modified," in 1718 he substituted the words: "the rays of light can still be disposed differently." And in the course of some extended remarks toward the close of this section, we find him writing:

We learn, further, that the colors we perceive after the refraction or reflection of light are not produced by some new modifications this

---

[156]*Experiences de physique, Par M. Pierre Poliniere, Docteur en Medecine* (Paris, 1709), p. 413.

[157]*Experiences de physique, par M. Pierre Poliniere, Docteur en Medecine. Seconde edition, revûë & beaucoup augmentée* (Paris, 1718), p. 455. A marginal note cites the Latin *Optice* of 1706.

light receives, and even that the whiteness of the sun's light is composed of all the principal colors, mingled together in some fashion. . . . Furthermore it is also to be noted that all light of the same kind has its characteristic color and refrangibility [*sa réfraction particuliere*], and that this color cannot be altered by any reflections or refractions.[158]

And then he concludes:

> For the rest I do not claim that these new ideas [*nouveautez*] are incontestable and without difficulty. I only hope that they will contribute to extending further that investigation to which we are invited every day by the objects we see around us.[159]

It is obvious, then, that the discussion of color in a popular book of experimental physics was, between the years 1709 and 1718, profoundly altered. During this interval, in 1712, Malebranche had come to accept Newton's description of his experiments and his revolutionary doctrine of color. In 1716 and even more clearly in 1717, Dortous de Mairan, under the influence of Malebranche, showed in his writings a familiarity with Newton's optical work and indeed repeated some experiments successfully. In 1715, as we have seen, a group of French academicians had watched Desaguliers, in London, repeat with success some of Newton's controversial experiments on light and color. Their return to Paris must have led to considerable discussion within the Academy and in the scientific world outside. Polinière would certainly have heard echoes of all this from his old teacher, Pierre Varignon, the mathematician and physicist and friend of Malebranche; living in Paris, he would certainly have kept in touch with Varignon. It is not too bold a conjecture to suggest that Varignon was important in Polinière's conversion.[160]

---

[158]Ibid., p. 463.

[159]Ibid., p. 465.

[160]In his second edition of the *Expériences de physique* Polinière cites Newton's *Optice* of 1706, to which Varignon and the current excitement may have led him. In later editions his reference is to Coste's translation of 1720.

VII

Not long after the French travelers' return to Paris, Newton's chief optical experiments were successfully repeated in the presence of various members of the Academy of Sciences. The details of these demonstrations (for there were in fact two) have not been preserved, and there is no reference to them in the *Histoire et Mémoires* of the Academy nor, so far as I have discovered, in any learned journal. Our chief informants are Pierre Coste, the translator of the *Opticks* into French and, more remote in time from the events, Montucla in his *Histoire des mathématiques*. Coste gives us slightly differing accounts in the two versions of his translator's preface.[161] From his preface in the Amsterdam edition of 1720 we learn that Jean Truchet, who as a Carmelite friar took the name of Père Sébastien, and who since 1699 was an honorary member of the Academy of Sciences, had "confirmed most of the experiments of [Newton's] Treatise on Colors" before a distinguished audience that included the Cardinal de Polignac, churchman, diplomat, and writer; Fontenelle, the *secrétaire perpetuel* of the Academy; and Pierre Varignon among others.[162] A letter of Père Sébastien to Sir Isaac Newton[163] makes an interesting, if romanticized, allusion to these experiments:

Scarcely [he writes] had that work of yours on light and colours come to my notice, when a deep and ardent eagerness entered my mind to

[161]For the publication of these two French editions see below, pp. 144–163.

[162]*Traité d'Optique*, 2 vols. (Amsterdam, 1720), I, xi–xii. These facts were reported in a review of Coste's translation that appeared in the *Journal des sçavans*, 67 (1720), 546. The list of observers is altered (corrected?) in the edition of 1722. Fontenelle's name is missing, but so is that of the Cardinal de Polignac, who was undoubtedly present. Instead we find the names of a "M. Jaugeon" and of "M. Jussieu." The latter is certainly Antoine de Jussieu (1686–1758), founder of a dynasty of botanists at the Academy and the Jardin du Roi. Jaugeon, whose date of birth and Christian name are unknown, was appointed a *pensionnaire mécanicien* in 1699 and died in 1724 as a *pensionnaire vétéran*. He did not rate a Fontenelle eulogy.

[163]Born Jean Truchet in 1657, he entered the Carmelite order at age seventeen, taking the name of Sébastien. Made an honorary member of the Academy in 1699, he resigned from that body in 1726. Fontenelle in his *éloge* (*Oeuvres*, VII, 308–321) stresses his skill in applied mechanics and hydraulics, but does not mention the experiments that concern us.

try those experiments which you bring together so brilliantly in your treatise. Without delay, taking as my guide and as it were protector an English copy of your treatise, and especially benefiting from the assistance of Mr. Geoffroy, whom it was proper and useful to employ as interpreter of your tongue, I tried the majority of the experiments cited by you in the presence and with the approval of the most eminent Cardinal de Polignac and also that of a very numerous group of aristocrats and men deeply versed in physics, who are, as it were, blood-brothers of yours.[164]

From this letter two interesting facts emerge. First, the experiments were performed at Père Sébastien's house which, he tells us, faced south and usually provided sunlight for seven or eight hours each day in the summer and three or four in the winter; and his room, forty-two feet long, allowed him to use a series of mirrors to compensate for the changing direction of the sun's rays and to intercept, separate, and refract the same ray of light, even several times over, before it reached the farthest point of the room. The second fact is that a M. Geoffroy, almost certainly Etienne François, our old friend who had expounded the contents of the English *Opticks* several years earlier before the Academy, helped to interpret the English text. Perhaps he made available to the experimenter his abridged and unpublished French version.[165]

A somewhat different perspective, notably as it concerns the enthusiasm Père Sébastien evinced in his letter, and the mere

[164]Sébastien Truchet [*sic*] to Newton, a Latin letter of 1722 or 1723 in Newton *Correspondence*, VII (1977), 111–113, and 113–118, no. 1350. The first number (or numbers) refers to the Latin original, the second to the English translation. Here I cite unmodified the editors' translation. The letter expresses Père Sébastien's thanks for a copy of "illius optimi et exquissitissimi de optica libri," presented in Newton's name by Varignon. I do not agree with the editors of the *Correspondence* (p. 116, n. 2) in their hesitancy as to which edition was presented. It was almost certainly the Paris edition of 1722, a handsome book that answers Père Sébastien's description. Varignon, who slaved over it, and was used by Newton to distribute copies to selected friends, had nothing to do with the Amsterdam *Traité d'Optique* of 1720.

[165]Newton *Correspondence*, VII, pp. 111 and 114. The description "cum maxime opera famosi inter nos academici domini Geoffroy" hardly describes the younger Geoffroy whom Hall took to be the man in question ("Newton in France," p. 244). In his notes to vol. VII of the Newton *Correspondence*, published two years after this paper, Hall took a newer view, and settled for Etienne François.

"approval" of Polignac, is provided by an unpublished letter of the great scientist Réaumur to an unknown recipient. The experiments on color that the Academy of Sciences was supposed to carry out, he writes, have been held up (*arestées*). "The Cardinal de Polignac who had his heart set on them charged Père Sébastien with preparing everything, and after all the preparations were made, Père Sébastien left for Lorraine."[166] It is possible that Réaumur's reference is not to the experiments Coste tells us were carried out by Père Sébastien in 1719 but to those performed somewhat later, probably in 1721, by Nicolas Gauger (1680–1730).[167]

Like his predecessor, Gauger was known to his contemporaries for his inventions and his mechanical dexterity,[168] and he may have been called upon to replace Père Sébastien at the last moment. A lawyer at the Paris Parlement, he is remembered today, if at all, as the author of an anonymous book on stoves, *La mécanique du feu* (1713), which was translated into English by Desaguliers with the latter's improvements and embellishments.[169] Again it was Polignac, according to Montucla, who inspired these second experiments and bore the cost of the excellent English prisms that were used. Montucla adds that, after the successful conclusion of the experiments, the Cardinal received a letter of appreciation from Newton.[170] We cannot be certain which set of experiments is referred to; in any case, if Newton did write such a letter it has not surfaced. Among the observers that Coste lists as present at this second demonstration, besides Polignac, was the chancellor of France, Henri François d'Aguesseau (1668–1751), chief law officer of the crown; his sons and a nephew; a M. de Lagny; and the

---

[166]This letter is excerpted at some length in no. 31810 of the Charavay Catalogue, no. 725 (July 1967). I am most grateful to the Maison Charavay for generously providing me with a photocopy of this letter. I believe the passage I have quoted refers to the second set of experiments, those carried out by Nicolas Gauger, probably in 1721. Gauger seems to have been called upon when Père Sébastien left for Lorraine. The editors of the Newton *Correspondence* take this position.

[167]*Traité d'Optique* (Paris, 1722), Préface du traducteur.

[168]Montucla, II, 626.

[169]For this book, its translation by Desaguliers, and its influence on Benjamin Franklin, see Cohen, *Franklin and Newton*, pp. 261–264.

[170]Montucla, II, 626.

Oratorian father, Charles René Reyneau (1656–1728), mathematician and devoted follower of Malebranche.[171]

Of Polignac's central role in these demonstrations there can be little doubt, but there is also a touch of irony.[172] A staunch Cartesian, Polignac is best remembered, apart from his stellar diplomatic career, for his *Anti-Lucretius, sive de deo et natura* (1745), a didactic poem on which he labored during his leisure hours and which was published posthumously by his friend, the Abbé Charles d'Orléans de Rothelin.[173] In this long poem he attacked the materialism of Epicurus and Lucretius, and of such of their modern followers as Pierre Gassendi; he disliked what he spoke of as Newton's "Epicureanism," although he praised his genius, and expressed wonder that such a great man (*tam doctus tamque sagax*) could deprive space of all matter, and instead of invoking contact forces could call upon magical entities like attraction. Back in Paris between diplomatic assignments, Polignac, who had a lifelong amateur interest in natural philosophy, entered the Academy of Sciences as *académicien honoraire*, replacing Father Malebranche who had just died. Despite his loyalty to the Cartesian world view, Polignac became interested in Newton's theory of light and color, perhaps as a result of Malebranche's conversion to it, perhaps persuaded by the reports of the academicians who had earlier returned from England in 1715. As Montucla put it, he "spared no expense" to have Newton's theory of color verified.[174] And according to Dortous de Mairan, who wrote his academic eulogy, the cardinal planned to incorporate an exposition of the new color theory into his *Anti-Lucretius,* but did not

[171]*Traité d'Optique* (Paris, 1722), Préface du traducteur. "M. de Lagny" was Thomas Fantet de Lagny (1660–1734), who entered the Academy in 1696 and replaced Varignon in 1723 as *pensionnaire géomètre* on the latter's death. For Père Reyneau, a disciple of Malebranche and an *associé libre* of the Academy, see Chapter 3, pp. 64–65.

[172]For Polignac, see the *éloge* by Dortous de Mairan, *Hist. Acad. Sci.* for 1741 (1744), pp. 180–200. A modern work is Pierre Paul, *Le cardinal Melchoir de Polignac* (Paris, 1922), esp. chap. 9, "L'écrivain, l'amateur, le courtisan, le diplomate," where the *Anti-Lucretius* is discussed.

[173]I have used the 12mo edition (two vols.-in-one) of 1749. The French translation, by Jean Pierre de Bougainville (Paris, 1749), the *secrétaire perpetuel* of the Académie des inscriptions et belles-lettres, is extremely free and more flattering toward Newton than the Latin original.

[174]Montucla, II, 626.

live to carry it out. Nevertheless, we find in that work a few lines in which Newton is praised for having "decomposed a ray of light" into seven primitive and permanent colors by passing it through a prism.[175]

Purely on its scientific merits, this second repetition in Paris of Newton's experiments must seem otiose and scarcely deserving of the space accorded it in Coste's revised preface. Some special reason may have occasioned it. Clearly, there must have been a connection between the presence of such a busy and illustrious official as the chancellor of the realm and the active support he was later to accord the project of a Paris edition of Coste's *Traité d'Optique*. Could it have been to arouse his interest that this second demonstration was organized? If so, it was eminently successful.

D'Aguesseau was born in Limoges into the *noblesse de robe*.[176] Educated largely by his father, the *intendant* of Limousin (and later of Languedoc), he became an attorney (*avocat du roi*) at the Châtelet, the famous tribunal of first instance in Paris, at the young age of twenty-two. As a gifted orator and precocious legal thinker he was clearly destined for the top ranks of his profession. In 1700 Louis XIV named him general prosecutor (*procureur général*) of the Parlement of Paris, a post he held until 1717 when the regent, Philippe d'Orléans, as a reward for d'Aguesseau's part in breaking the will of the Sun King, raised him to the rank of Chancellor, thus making him the second most powerful personage in France. Except for two brief exiles to his country estate, he held this high office, "the mouth of the Prince, interpreter of his wishes," until his death in 1751. An adroit manipulator, strongly Gallican in his religious poli-

[175]"Et radium solis transverso prismate fractum Septem in primogenos permansurosque colores [solvit]." *Anti-Lucretius, sive de deo et natura*, 2 vols. (Paris, 1749), I, 70.

[176]For d'Aguesseau, who preferred the spelling Daguesseau to avoid stressing the particle, see Grandjean de Fouchy's *éloge* of d'Aguesseau in *Hist. Acad. Sci.* for 1751 (1755), pp. 178–194, and Francis Monnier, *Le chancellier d'Aguesseau* (Paris, 1860). More recent, and with a good bibliography, is Georges Frêche, *Un chancellier gallican: Daguesseau* (Paris, 1969). A good introduction to the world of the *noblesse de robe* is Franklin L. Ford, *Robe and Sword* (Cambridge, Mass., 1962). For the great jurist's contact with Father Malebranche, see *Oeuvres de Malebranche*, XX (1967), 203–204, and Frêche, p. 9. Both are derived from Henri François d'Aguesseau, *Oeuvres complètes*, 16 vols. (Paris, 1819), XV, 31.

cies, he ranks as the outstanding legal reformer in eighteenth-century France, striving for uniformity in legal principles, simplification of the judicial process, and a codification of the civil law. In attempting to bring order into the mare's nest of provincial customary laws, he achieved some lasting success: certain of his reforms and consolidations made their way into the great *Code Napoléon.*

That an official of this eminence was persuaded to attend Gauger's demonstration of Newton's optical experiments is less surprising if one recalls the extent of d'Aguesseau's education and the breadth of his interests. Besides a linguistic range rare in his time and place—a solid foundation in Greek and Latin, of course, and some reading knowledge of Hebrew and several modern languages including English—he had learned something of natural philosophy (*la physique*) and astronomy. He was profoundly influenced by Descartes whose writings he warmly recommended to young lawyers. Scientific works occupied an important place in his library, and he was particularly attracted by mathematics. This interest, like his religious convictions, was strongly reinforced by reading Malebranche, and, so it is said, by study and discussions with the great Oratorian philosopher himself. We have already noted his contact with Malebranchistes like the mathematician Père Reyneau, who was in attendance at Gauger's performance, and with Polignac. It is hardly to be wondered at that, in the event, he was made—like Malebranche himself and Polignac—a *membre honoraire* of the Academy of Sciences.

VIII

The final step in the naturalization of Newton's optical discoveries in France was the publication, not long after the events we have described, of Pierre Coste's French translation of the *Opticks,* the first translation of that work in any modern language. Made it seems from the second edition of 1717/18, and entitled *Traité d'Optique,* it was first published in Amsterdam in two duodecimo volumes, with a bicolored title page, but otherwise undistinguished in typography and appearance. It had a preface by the translator,

Plate 2. Henri François d'Aguesseau (1668–1751), Chancellor of France in 1720–1722. Frontispiece to Vol. I of d'Aguesseau's *Oeuvres complètes* (Paris, 1819).

and a surprising conclusion that does not appear in any of the English editions.

Like Desaguliers, Pierre Coste (1668–1747) was a Huguenot refugee. Born in Uzes, in the old province of Languedoc, he was forced to leave France by the revocation of the Edict of Nantes and the persecutions that ensued.[177] He would have been in his late 'teens when he betook himself to Amsterdam where, at a synod held in 1690, he was admitted to the Protestant ministry. He rarely preached, and soon gave up pastoral duties for works of translation and criticism, taking advantage of his considerable linguistic ability and keen literary interests. He prepared, with commentaries, editions of Montaigne's *Essais,* the *Fables* of La Fontaine, and the *Caractères* of Theophrastus and La Bruyère. An early—and his most original—contribution was his *Histoire de Condé* (1693), a work that went through several editions and was translated into English.[178]

We do not know when, precisely, Coste left for England; but in 1697, while at work on a French translation of Locke's *Essay concerning Human Understanding,* he became tutor to the son of Sir Francis and Lady Masham and took up residence at Oates in Essex. Here Locke, an honored guest at Oates in his declining years, took a keen interest in Coste's translation and is said to have superintended it. It exerted a major influence, far greater, needless to say, than his translation of Lady Masham's *Discourse on Divine Love,* greater even than the other works of Locke or even Newton's *Opticks,* in grafting an English element upon the French Enlightenment.

Pierre Coste gives little evidence of scientific tastes; his only translation of a scientific work was a rendering in 1708 of Francesco Redi's important parasitological treatise from Italian into Latin.[179] How he came to make his French version of Newton's

---

[177]For Coste see the article by Frank T. Marzials, in the *Dictionary of National Biography,* which is superior to that in the Didot-Hoefer *Nouvelle biographie générale.*

[178]*Histoire de Louis de Bourbon, IIe du nom, prince de Condé* (Cologne, 1693). The subject, of course, is the victorious soldier of the reign of Louis XIV whom the French have called "le grand Condé."

[179]This was Redi's *Osservazioni intorno agli animali viventi che si trovano negli animali viventi* (Florence, 1684). If not the Newtonian translation, one wonders what justified Coste's election as F.R.S. in 1742, shortly before his death.

Opticks is only a matter for conjecture. In his translator's preface he tells us that he undertook the translation "par l'ordre d'une grande Princesse," and that he obtained Newton's consent to prepare and publish it.[180] He acknowledges, too, the assistance of Desaguliers "who had the kindness to look over my manuscript with great care." Desaguliers may have suggested the project, although that is doubtful. Coste is at pains to recall that in London Desaguliers had shown the key experiments of the *Opticks* to M. Rémond de Montmort, M. le Chevalier de Louville, and other members of the Academy of Sciences. Possibly one of this group, or someone like Varignon, who saw the experiments repeated in France in 1719, suggested the idea of a translation into French. Another possibility is Fontenelle, whom Coste mentions as having observed the first repetition, and who is accorded a gracious compliment: Coste hoped he could combine exactness of translation with "ce tour vif et délicat" that is the hallmark of the perpetual secretary of the Academy of Sciences.[181]

At all events, this first French edition appeared early in 1720. A copy had reached Paris and been gone over by Varignon, before the end of April, and the translation was reviewed in the May issue of the *Journal des sçavans*, where the reviewer commends Coste for having made the translation, since few persons in France were "instruites à fond des raisons dont se sert M. Newton pour soutenir son système."[182]

If Varignon was not involved in the first French version, he played a central role in the revised second edition, published in Paris in 1722. A mathematician, hardly in the top class, Varignon had published in 1687—the same year, in the interest of contrast, that Newton's *Principia* appeared—his *Projet d'une nouvelle mécanique,*

---

[180]*Traité d'Optique* (Amsterdam, 1720), I, Préface du traducteur, iii. The "grande Princesse" was that patroness of scholars, Caroline of Anspach, princess of Wales, who instigated the famous Leibniz-Clarke correspondence.

[181]Hall remarks that Coste was probably mistaken when he listed Fontenelle as one of the observers of Père Sébastien's experiments. See his "Newton in France," p. 24, and Newton *Correspondence*, VII, 117, n. 6. Fontenelle's name is removed from the list in the Paris edition of the *Traité d'Optique* (1722), although the compliment remains; it may be significant, too, that Fontenelle makes no mention of the experiments in his *éloge* of Père Sébastien.

[182]Jacqueline de la Harpe, p. 322.

which brought him admission to the Academy of Sciences the following year and an appointment as professor at the newly founded Collège Mazarin. Here he taught, and indeed lived, for the rest of his career, adding to his responsibilities a chair at the Collège Royal.[183] Influenced by Malebranche and his circle, he came to accept the Leibnizian calculus and was one of the first in France to apply it in cautious fashion to problems of Newtonian dynamics.[184] When the scientific world was split by the heated debate over the invention of the calculus, Varignon strove to remain impartial, even trying to act as peacemaker between Leibniz's disciple Johann Bernoulli and Newton. Although, so far as we know, Varignon never visited England, he was soon in communication with English scientists, notably the mathematician Abraham De Moivre. On 24 November 1713, Varignon wrote to Newton—it seems to have been his first contact with the great man—thanking him for a gift of the second edition of the *Principia*. Already, he wrote Newton, he had asked De Moivre to procure it for him as soon as it appeared "having no reason at all to presume that you would give it to an unknown person."[185] In July of the following year Varignon was elected to the Royal Society;[186] and from a letter to Newton, expressing his appreciation of the honor, we are told that it was through Newton's influential patronage and nomination that he was chosen.[187] This was not the end, but only the beginning, of an exchange of letters, a number of which dealt with the stand taken by Johann Bernoulli in the calculus quarrel. Upon his election to the Royal Society, Varignon kept Newton abreast of scientific de-

---

[183]See Fontenelle, "Eloge de Varignon," *Oeuvres de Fontenelle*, VII, 146–162. Pierre Costabel's sketch in DSB, XIII (1976), 584–587, stresses Varignon's work in mechanics, but tells us little or nothing about his professional life or his relationships with foreign men of science. Cf. Brewster, II (1855), 73, and Newton *Correspondence*, VII, passim, the latter unavailable to Costabel when he wrote his sketch.

[184]See Chapter 3, pp. 57 and 60–61.

[185]Newton *Correspondence*, VI, 41 and 43.

[186]Journal Book, Royal Society of London, for 29 July 1714 (O.S.). Elected on the same day were Desaguliers and Martin Folkes, later president of the Society.

[187]For Varignon's letter of appreciation (7 November 1714), delayed by his three-months' absence from Paris in the country, see Newton *Correspondence*, VI, 187 and 187–188, no. 115.

Plate 3. Pierre Varignon (1654–1722).

velopments in France by sending him (as indeed was Newton's due as a foreign member of the Academy of Sciences) the successive volumes of the Academy's *Histoire et Mémoires* and of the *Connoissance des temps,* an almanac published under the Academy's auspices. In the summer of 1718, Newton reciprocated by sending Varignon three copies of the second English edition of his *Opticks* (the edition, as we saw, from which the French translation was made) for distribution among Varignon's colleagues and friends.[188] Later in the year Newton dispatched five copies of the second Latin *Optice* (1719), one intended for Varignon, others for Fontenelle, Montmort, the library of the Academy, "and a fifth for any other friend who understands optical matters."[189] The correspondence continued, the mutual esteem deepened, and soon Newton and Varignon exchanged portraits. The portrait of Newton destined for Varignon was painted by Sir Godfrey Kneller. A reproduction appears as the frontispiece of the final volume of *The Correspondence of Isaac Newton* VII (1977).

For Varignon's role in shepherding through the press a revision of Coste's earlier translation of the *Opticks,* there is ample evidence in the concluding volume of the Newton *Correspondence.* But whether Varignon instigated the publication of the handsome Paris edition of 1722, we can only wonder. In 1719, the year in which the experiments of Père Sébastien were carried out, Varignon, whom we know to have been in the audience, had risen to high rank in the Academy of Sciences; indeed, in that year he was its *directeur,* or chief scientific officer. Soon after Coste's translation had appeared in Amsterdam, the *garde des sceaux,* the high official responsible for approving the publication of books, charged Varignon with determining whether it deserved publication in France with royal

[188]Brewster, II, 73. The letter of Newton mentioning these complimentary copies is in Newton *Correspondence,* VII, 2–4, no. 1298. In Varignon's letter of acknowledgment, dated 17 November 1718, he remarks that he has presented a copy on Newton's behalf to the Abbé Bignon, and one to Johann Bernoulli "in order that I might reveal your generous nature to him." The third, inscribed to him in Newton's own hand (not, indeed, Newton's usual practice), Varignon kept.

[189]Newton *Correspondence,* VII, 5–6, no. 1300. Again, Varignon sent a copy as a peace offering to Johann Bernoulli on Newton's behalf "in the same spirit" in which he had sent the copy of the second English edition.

approval. Varignon's *approbation,* filled with praise for the book and its author, was submitted on 28 April 1720.[190] Not long after, probably sometime in May, Varignon informed Newton, in the postscript to a letter, that one of the booksellers of Paris was seeking the official *privilège* of publishing "your excellent book of *Opticks* as translated into French by M. Coste in Holland," and that he, Varignon, had approved it.[191] Whether the bookseller, François Montalant, had suggested the project, or merely been recommended to Varignon, is hard to decide, but the second possibility is more likely in view of Montalant's subsequent performance.

We have lost, it would seem, an important letter of Varignon to Newton describing in some detail the negotiations with Montalant, the costs of publication and similar matters, but its contents can be roughly reconstructed from Newton's reply, seemingly written early in August 1721.[192] Clearly, from the time that Varignon approved granting the *privilège,* the publisher made difficulties, and the negotiations dragged on with no end in sight. Varignon resolved to bring out his heavy artillery. With an unidentified friend of Père Reyneau as intermediary, an appeal was made to the Chancellor. D'Aguesseau used his ample authority to bring Montalant to heel, and a contract was finally agreed upon.[193]

[190]This *approbation* and Coste's (revised) translator's preface, both of which graced the Paris edition, have been regrettably omitted from the modern facsimile published by Gauthier-Villars in 1955. For the role of the *garde des sceaux* in the business of book censorship, see Newton *Correspondence,* VII, 91–92, n. 8. I believe the editors are mistaken when they identify the holder of this office as Fleuriau d'Armenonville. The *Almanach Royal* for 1720 lists the Marquis d'Argenson as the holder of this post. See also Marcel Marion in his *Dictionnaire des institutions de la France aux XVII*ᵉ *et XVIII*ᵉ *siècles* (Paris, 1923), who gives the Marquis d'Argenson the title for 1718–1720 and Fleuriau d'Armenonville (secrétaire d'Etat de la guerre) for 1722–1727.

[191]Newton *Correspondence,* VII, 90 and 91, no. 1338.

[192]Newton to Varignon, undated, ibid., pp. 141 and 142–143, no. 1363. The editors give the date as "early August 1721."

[193]Newton wrote to Varignon (enclosing a letter to d'Aguesseau) thanking him for the "favours with which he has honoured me") and remarking: "I have considered your contract with the bookseller, and instead of twelve pounds sterling am sending twenty. And this sum may be paid whenever [you] like, and the costs of binding the books to be given to friends can be taken out of the cash transferred." He also asked that his thanks be conveyed to Father Reyneau and "also to the friend by whose intervention the business was brought to the attention of the Chancellor" (ibid., pp. 141 and 142, no. 1363). Editors' translation.

For some time Varignon had been engaged in the task, which was to occupy him for many months, of approving and embodying (probably by entering them in a master copy of the Amsterdam edition) the textual changes in Coste's *Traité d'Optique* that came to him from England. The primary revisions were made by Varignon's friend, Abraham De Moivre, submitted for Newton's approval, and then sent on to Paris.[194]

As for Coste, he seems at first to have been virtually ignored. He had informed Varignon of some corrections he proposed to make, but Varignon had replied that Coste should consult Newton about his corrections, stating that he would not introduce any changes without Newton's permission. Coste felt ill used, for only a few of De Moivre's corrections had been shown him. He wrote Newton plaintively asking that his own changes and his "reflections" on De Moivre's be forwarded to Paris, as Newton said he would do; and Coste promised that in his revised translator's preface he would note the pains that De Moivre had taken to improve the new edition.[195] Newton seems to have been sympathetic and friendly toward Coste, and in fact did send the latter's corrections to Varignon, who was obviously less than happy to receive them.[196]

With the contract signed, the printing began on 1 August 1721,[197] and a sample sheet was shown to d'Aguesseau, who found the type font, obviously old and worn, totally unacceptable. In September, Varignon described to Newton how he had paid a visit to the chancellor, bearing a new specimen of the first sheet, printed with a

[194]Since the overwhelming number of changes concerned French style, Newton could hardly have exerted much influence, for he read French, as he once confessed, only with difficulty and the help of a dictionary.

[195]Pierre Coste to Newton, 16 August 1721 (O.S.), in Newton *Correspondence*, VII, 147–148, no. 1365. Coste was as good as his word.

[196]Varignon to Newton, 18 September 1721, ibid., pp. 154 and 155–156, no. 1369). Newton thanked Varignon profusely "because you have been so good as to take upon yourself the task of comparing together the corrections of Mr. Coste and of Mr. De Moivre, and choosing what shall seem the best. . . . I was quite fearful that Mr. Coste's corrections would make too much trouble for you, busy as you are with many other things. But since you were so kind as to take on this task, you put me more heavily in your debt" (Newton to Varignon, 26 September 1721 [O.S.], ibid., pp. 160 and 163, no. 1372). Editors' translation.

[197]Varignon to Newton, 4 August 1722, ibid., pp. 206 and 207–208, no. 1396.

better font. This, in turn, d'Aguesseau rejected. Once again the beleaguered and "evasive" publisher was brought before the chancellor, who commanded Montalant in what Varignon called a "rather imperious tone" (*altior vox*) to try again with an absolutely unblemished font.[198]

This tactic evidently worked, at least for a time, because the first sheet was finally approved by d'Aguesseau, and Varignon in late September sent it on for Newton's inspection, deploring the fact that the whole month of August and part of September had been wasted on account of all the "evasions" of Montalant. He promised nevertheless to send the remaining sheets as soon as this "sluggish" fellow supplied them.[199] Although he appears to "be drowsing" over his work, the bookseller is under steady pressure from friends, "armed with the authority of our illustrious chancellor" to push the business forward diligently. Newton was pleased with the typography, describing it as "elegant and of noble appearance,"[200] and by the end of November he had received the sheets completing Part I of Book I of the Paris *Traité d'Optique*. Others, Varignon promised, were to follow, but exactly when he could not foresee "as the printing proceeds extremely slowly, although I often complain about this sluggishness."[201]

Early in 1722 Varignon was obliged to report that the printing had slowed down and nearly stopped, but that the chancellor's brother had threatened Montalant with the loss of his *privilège* and thus had stirred up the "faithless bookseller" to greater efforts so that matters proceeded much more rapidly than before.[202] But the fall of the chancellor gave the bookseller, as he thought, a reprieve, and the work came to a virtual standstill. "Then," wrote Varignon

[198]Varignon to Newton, 18 September 1721, ibid., pp. 152–154, and 154–156, no. 1369. In his *éloge* of d'Aguesseau (see n. 176), Grandjean de Fouchy stressed the Chancellor's artistic sensibility: "il apprit à dessiner, &, ce qui étoit bien plus considerable, à connoître les beautés de la peinture & la main des meilleurs maîtres" (p. 180).

[199]Newton *Correspondence*, VII, 152–154 and 154–156, no. 1369.

[200]Newton to Varignon, 26 September 1721 (O.S.), ibid., pp. 160 and 163, no. 1372. Newton was more sympathetic toward Montalant than his French collaborators. See n. 193.

[201]Varignon to Newton, 9 December 1721, ibid., pp. 178–180, no. 1381.

[202]Varignon to Newton, 13 January 1722, ibid., pp. 183–184, no. 1385.

on 4 April 1722, "stimulated by this last wickedness [*ultima nequitia*] of Montalant, I wrote an angry letter, not to that shamless trickster [*inverecundus fallacus*], but to the printer," extracting the promise of an effort to speed things up. For the moment this seemed to be working wonders.[203]

In early August 1722 Varignon conveyed to Newton, with his usual fulminations against Montalant, the news that he was sending the final sheets of the text of the *Traité d'Optique,* together with specimens of the ornamental capital letters.[204] After giving his opinion of the relative value of the changes proposed by De Moivre and by Coste he wrote:

> For the rest, where I have seen him [Coste] differing in any way from De Moivre, which was rare, I have considered with as much care and fairness as I could the opinion of each man, and if I still remained undecided I consulted the most recent Latin edition on this point; I would have consulted the [second] English, if I had understood that language.[205]

The work was completed on the last day of July 1722, and specially bound copies were distributed by Varignon to friends on the Continent. It is a handsome quarto with folding plates. Each major division or "Part" is illustrated by a headpiece or vignette showing an idealized, perhaps in a sense false, version of the famous *experimentum crucis* with two prisms, employing a lens of which there was no mention by Newton in his first paper nor where that experiment is described in the *Opticks.* The first word of each Part of Books I and II, and the opening word of Book III, which is not divided into Parts,[206] all begin with decorative initial letters, gener-

---

[203]Varignon to Newton, 4 April 1722, ibid., pp. 193–194, no. 1390.

[204]Varignon to Newton, 4 August 1722, ibid., pp. 206–208, no. 1396.

[205]Ibid., p. 206. Varignon found certain corrections by Coste helpful, but De Moivre's more so, for they dealt with more places in the book and often with the mathematical sections where Coste was less qualified to comment than De Moivre.

[206]The puzzling "Part I" of Book III, which makes its appearance in the second English edition of 1717/18 and is copied uncritically in the second Latin of 1719, the third English of 1721, and the posthumous fourth English edition of 1730, is quietly dropped from the Amsterdam *Traité d'Optique* in 1720 and the revised French translation of 1722. For the original purpose of this "Part I" see "Newton's Optical Aether: His Draft of a Proposed Addition to His *Opticks*," in my *Essays and Papers*, pp. 120–130.

TRAITÉ
D'OPTIQUE
SUR

LES REFLEXIONS, REFRACTIONS,
INFLEXIONS, ET LES COULEURS,
DE
LA LUMIERE.

Par Monfieur Le Chevalier NEWTON.

Traduit par M. COSTE , fur la feconde Edition Angloife,
augmentée par l'Auteur.

SECONDE EDITION FRANCOISE,
beaucoup plus correcte que la premiere.

A PARIS,
Chez MONTALANT , Quay des Auguftins , du côté
du Pont faint Michel.

M. DCC. XXII.
AVEC APPROBATION ET PRIVILEGE DU ROY.

Plate 4. Title page of the second French translation of Newton's *Opticks.*

ally enlivened with putti, some supposed to be observing optical phenomena. No decorative elements of comparable quality had been used in any of the London editions, English or Latin, or in Coste's Amsterdam translation. Without question this is the handsomest edition of the *Opticks* printed in Newton's lifetime, or indeed later.

The idea for a vignette arose, predictably enough, in Paris. When Varignon, early in September 1721, sent Newton the first sheet of the new French edition for his approval, a blank space had been left above the half title for an engraved headpiece or vignette which was to adorn each Part of the work. Varignon called this to Newton's attention, and asked him to "think of an idea" appropriate to his book, adding that Newton would "help us greatly" if he would forward a drawing prepared by a skilled English artist to be engraved in Paris.[207] Newton approved the idea, came up with the subject for the vignette, and intimated to Varignon that he had sought out an artist in England to render his sketch professionally "but it is not yet drawn by the artist. I will spur him on."[208] His first choice may well have been Jacques Antoine Arlaud (1688–1743), a Swiss artist who was in London in the autumn of 1721. Yet Arlaud was not, as the editors of the Newton *Correspondence* would have us believe, the artist who executed the vignette for the engraver.[209]

A native of Geneva, who had come to Paris at the age of twenty and where he earned a substantial reputation, Arlaud arrived in England with a letter of introduction to Caroline of Anspach, the princess of Wales, whose portrait he is said to have painted. During this visit Arlaud made Newton's acquaintance.[210] The relationship

[207]Varignon to Newton, 18 September 1721, in Newton *Correspondence*, VII, 153 and 155, no. 1369.
[208]Newton to Varignon, 26 September 1721 (O.S.), ibid., pp. 160 and 163, no. 1372.
[209]Ibid., p. 165, n. 10, and pp. 213–214, n. 1, where the editors write that Arlaud "whilst in England . . . drafted the vignette for the second edition of the *Traité d'Optique*. . . . He was apparently also responsible for drawing the figures." I take this to mean the folding plates. In any case, I can find no evidence to support either statement.
[210]For Arlaud see J. B. Descamps, *La vie des peintres flamands, allemands et hollandois*, 4 vols. (Paris 1753–1763), IV, 116–122, cited by the editors of the Newton *Correspondence*; also Albert de Montet, *Dictionnaire biographique des genevois et des vaudois*, 2

with Newton, mentioned in the chief biographical sketches, is confirmed by a letter of Varignon to Newton dated from Paris on 9 December 1721. "Mr. Arlaud," he wrote, "brought me most welcome news of you," and the two men conversed about their eminent English friend "most respectfully late into the night."[211] But there is no hint that Arlaud was thought of as the artist selected by Newton. Rather the contrary: Varignon tells Newton that he has received from De Moivre the "figure you have devised and sketched as an ornament for the beginning of the book"; nothing, he went on, could be more appropriate, and he adds that he is looking "for an ingenious and skilled draughtsman" to deploy properly the observers—including, as we learn elsewhere, the inevitable putti—to be added to the little picture (*tabella*).[212]

The picture actually received by Varignon is certainly not the rough drawing reproduced as Plate III in the first volume of the Newton *Correspondence*. Even as reproduced there it displays the earmarks of a crude preliminary attempt, not such as to be handed on to an artist to redraw (see Figure 2). There is indecision, shown by crossings out, as to where the beam of light should fall on the perforated board; indecision too as to how to place the motto: *Nec variat lux fracta colorem*. Even a casual look at the original in the Bodleian Library, Oxford, shows that Newton used not a virginal bit of paper, but the blank space on a small sheet previously used for other jottings, above and below the drawing, and on the verso of the sheet.

The sketch that was actually sent by Newton to Varignon, via De Moivre, is neat and decisive (see Figure 3). It is now in the possession of the Bibliothèque Publique et Universitaire de Genève. About five inches by three, its verso was originally blank, but now has an inscription, attributable to Arlaud (for it accompanies New-

---

vols. (Lausanne, 1877–1878), I, 13; the *Dictionnaire historique et biographique de la Suisse*, 7 vols., and supplement (Neuchatel, 1921–1934), I, 405; and Thieme and Becker, 37 vols. (Leipzig, 1907–1950). *Allgemeines Lexicon der bildenden Künstler*, II (1908), 110.

[211]Varignon to Newton, 28 November 1721, in Newton *Correspondence*, VII, 178–179, no. 1381.

[212]Ibid. The editors of the *Correspondence* translate "tabella" as "plate."

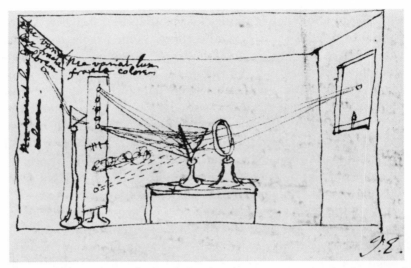

Figure 2. Newton's rough sketch for the vignette appearing in the second French translation of the *Opticks*. Courtesy of the Bodleian Library, Oxford.

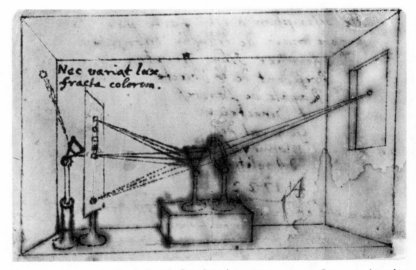

Figure 3. Newton's final sketch for the vignette as sent to Jacques Antoine Arlaud. Courtesy of the Bibliothèque Publique et Universitaire de Genève.

ton's letters of thanks, dated 14 September 1722, [O.S.]), reading: "Cette description a esté faite par la propre main de l'illustre Monsieur le Chevalier Isaac Newton President de la Société Royale de Londres."[213]

What then was Arlaud's role? The letters in Volume VII of the Newton *Correspondence* do not confirm the editors' assertion; instead they make the part he played quite clear. In his letter of 28 April 1722 Varignon wrote that he had searched repeatedly for an artist to draw the vignette and that he had found no one better to help him than M. Arlaud, "both by reason of the great skill he possesses in this art, and by the respect with which he honours you most highly."[214] Several days ago, he continued, he went to Arlaud for advice and help. Arlaud expressed his pleasure at the chance of being of service to Newton and agreed "to take the whole care and responsibility upon himself." Varignon accordingly showed Arlaud Newton's sketch, which the artist examined carefully; and since Arlaud was far better acquainted with the skilled draftsmen and engravers of Paris, Varignon charged him with selecting those whom he was to supervise to do the work. He was also to negotiate their fees. From this it is clear that Arlaud must be described as the artistic editor or supervisor of the Paris edition, but not strictly speaking the artist whose pen gave Newton's sketch its final form.[215] The artist Arlaud selected—his name appears on the vignette—was Jean Chaufourier (1679–1757) and the engraver chosen was Antoine Herisset (ca. 1685–1769).

Chaufourier, a native of Paris, was a painter (one authority calls him a *peintre-graveur*) who, at a later date, became a teacher of perspective at the *Académie royale de peinture et de sculpture* and is said to have taught drawing to the celebrated collector and critic,

---

[213]Bibliothèque Municipale et Universitaire de Genève, MS lat. 136. For the letter see n. 220.

[214]Varignon to Newton, 28 April 1722, in Newton *Correspondence*, VII, 199 and 200, no. 1391.

[215]Varignon wrote Newton that the specimens of capital letters (see above, pp. 154 and 156) had been "drawn and engraved *under the guidance and management* of Mr. Arlaud" (ibid., pp. 206 and 208, no. 1396). My italics.

Pierre Jean Mariette.[216] At the time he was chosen to draw the vignette for the *Traité d'Optique*, he was doubtless already at work on drawings for Jacques Bouillart's monumental *Histoire de l'abbaye royale de Saint-Germain des Prez* (Paris, 1724).

Herisset, who specialized in architectural subjects, had already earned a substantial reputation for the drawings and engravings he made for J.-B. de Monicart's sumptuously illustrated *Versailles immortalisé*, published in two volumes in 1720–1721.[217] Much later he was responsible for some twenty plates in the *Description de Paris* of Piganiol de la Force.[218]

With the letter of 28 April 1722, Varignon sent along Chaufourier's drawing to Newton "made ready for the engraver, if you think well of it." And he goes on:

> The rays which you drew in the original sketch, and which you accordingly see drawn in the copy, should be deleted in the opinion of Mr. Arlaud, and in mine also. For in the illuminated space nothing ought to be separately discerned, nor indeed can be; whence the illuminated space ought to appear wholly white. Moreover, he thought that the spectators and cherubs with which we had planned to adorn this drawing should likewise be omitted, especially as whatever figure is placed in the dark and shadowy part of the room must escape the sharpest sight, or at least be seen only as wholly black, which would be too gloomy. We now await your opinion of this drawing.[219]

A glance at the final product (see Figure 4) shows that Newton assented to the recommended changes, and that Chaufourier went back to the drawing board. Since the taste of the age ran to putti

---

[216]For Jean Chaufourier see E. Bellier de la Chavignerie and Louis Auvray, *Dictionnaire général des artistes de l'école française*, 2 vols. (Paris, 1882–1885), I, 243, where the name is misspelled "Chaufourrier," and Theime and Becker, *Allgemeines Lexicon*, VI (1912), 436, from which the sketch in J. Balteau et al., *Dictionnaire de biographie française*, VIII (Paris, 1959), cols. 842–843, seems ultimately to have been derived.

[217]Jean-Baptiste de Monicart, *Versailles immortalisé par les merveilles parlantes des bâtiments, jardins, bosquets, parcs, statues dudit chateau de Versailles*, etc., 2 vols. (Paris, 1720–1721). For Herisset, see Charles le Blanc, *Manuel de l'amateur d'estampes*, 2 vols. (Paris, 1854), I, 355, col. 1.

[218]Piganiol de la Force, *Description de Paris, de Versailles, de Marly, de St. Cloud, de Fontainebleau, et de toutes les autres belles maisons & de châteaux des environs de Paris*, 8 vols. (Paris, 1742).

[219]Newton *Correspondence*, VII, 199–200 and 201, no. 1391. Editors' translation.

TRAITÉ
D'OPTIQUE,
SUR
LA LUMIERE
ET LES COULEURS.

LIVRE PREMIER.
PREMIERE PARTIE.

ON deffein dans cet Ouvrage, n'eft pas d'expliquer les proprie-tés de la Lumiere par des Hypo-thefes ; mais de les expofer nuë-ment pour les prouver par le rai-fonnement, & par des Experiences. Dans cette vûë je vas commencer par propofer les Défini-tions & les Axiomes fuivants.

A

Figure 4. The vignette from the *Traité d'Optique* (Paris, 1722).

and rather crowded illustrations, we can be thankful that the good sense of Arlaud prevailed, and that we were spared more cherubs. We have, instead, a vignette which, if not scientifically precise, is straightforward and pleasing. This was certainly Newton's opinion, for in a letter written in the autumn of 1722, he expressed his thanks to Arlaud for having "corrected the diagram [*schema*] of the experiment in which light is separated into its primitive and immutable colours," and having made it far more elegant than before.[220] He had already expressed his debt in another fashion. The Bibliothèque Publique et Universitaire de Genève possesses, in the Salle Senebier, a sumptuously bound copy of the Paris *Traité d'Optique* accompanied by the following explanatory note:

> Ce livre relié a esté donné par l'illustre Autheur Monsieur le Chevalier Newton President de la Société Royale de Londres, et Directeur General de la monnoye d'Angleterre à Jacques Antoine Arlaud Citoyen de Genève par les mains de Monsieur Varignon Professor en mathematique au College Royal. A Paris le 14eme September 1722.[221]

We have now gone as far as planned for this study. The appearance of these two versions of Coste's translation of the *Opticks* may be said to bring our inquiry to an end, for it led to a general, or nearly general, acceptance in France of Newton's experiments and theory of color. Fontenelle had, like Arlaud and others, received through Varignon a specially bound complimentary copy of the *Traité d'Optique,* as well as a copy intended for the Academy of Sciences, and acknowledged both in a letter to Newton.[222] The capstone on this part of the Newtonian edifice in France was, as we saw

---

[220]Ibid., pp. 212–213, no. 1400. Editors' translation. First published in *Mémoires et documents publiés par la Société d'histoire et d'archaeologie de Genève*, V (Geneva and Paris, 1857), p. 366, and by J. Edleston in *Correspondence of Sir Isaac Newton and Professor Cotes* (London, 1850), pp. 188–189, where Edleston calls the recipient "Arland."

[221]The book and the note were examined by the present writer in the Bibliothèque Publique et Universitaire de Genève in early June 1979.

[222]Fontenelle to Newton, 22 November 1722, in Newton *Correspondence*, VII, 216, no. 1402. To the formal note acknowledging, as *secrétaire perpetuel*, the gift to the

at the beginning of this paper, Fontenelle's famous *éloge* of Newton. There were indeed doubters still, men such as Father Bertrand Castel, to whom Montucla applied the injunction of Xenocrates: *Apage, apage, ansas philosophias non habes.* Perhaps we may take the Abbé Nollet (1700–1770), the most successful public lecturer in physics in the French eighteenth century, as evidence of the final victory. Nollet's popular *Leçons de physique expérimentale* gave an important place to Newton's experiments on color, which he had repeated successfully as early as 1735.

Academy of Sciences, Fontenelle added a personal footnote thanking Newton for the copy sent to him and expressing delight at the praise bestowed upon him in the translator's preface. The remarks of Coste had already appeared in the Amsterdam *Traité d'Optique* (1720), which Fontenelle appears not to have read or even seen.

Index

Newton on the Continent

Designed by Richard E. Rosenbaum.
Composed by Eastern Graphics
in 10 point Baskerville, 3 points leaded,
with display lines in Baskerville.
Printed offset by Thomson/Shore, Inc. on
Warren's Olde Style, 60 pound basis.
Bound by John H. Dekker & Sons, Inc.
in Holliston book cloth
and stamped in Kurz-Hastings foil.

Library of Congress Cataloging in Publication Data

Guerlac, Henry.
    Newton on the Continent.

    English and French.
    Includes index.
    1. Newton, Isaac, Sir, 1642–1727—Addresses, essays, lectures. 2. Sci-
ence—History—Addresses, essays, lectures. 3. Physicists—Great Britain—
Biography—Addresses, essays, lectures. I. Title.
QC16.N7G82   509'.2'4 [B]   81-3187
ISBN 0-8014-1409-1   AACR2